Student Solutions Manual and Study Guide

FOR

SERWAY, VUILLE, AND FAUGHN'S

COLLEGE PHYSICS

EIGHTH EDITION, VOLUME 2

John R. Gordon
Emeritus, James Madison University

Charles Teague
Emeritus, Eastern Kentucky University

BROOKS/COLE
CENGAGE Learning™

Australia • Brazil • Japan • Korea • Mexico • Singapore • Spain • United Kingdom • United States

For product information and technology assistance, contact us at **Cengage Learning Academic Resource Center, 1-800-423-0563**

For permission to use material from this text or product, submit all requests online at **www.cengage.com/permissions** Further permissions questions can be emailed to **permissionrequest@cengage.com**

Cover Image: © Matt Hoover, www.matthoover.com

ISBN-13: 978-0-495-55612-1

ISBN-10: 0-495-55612-2

Brooks/Cole
10 Davis Drive
Belmont, CA 94002-3098
USA

Cengage Learning products are represented in Canada by Nelson Education, Ltd.

For your course and learning solutions, visit **academic.cengage.com**

Purchase any of our products at your local college store or at our preferred online store **www.ichapters.com**

Printed in the United States of America
1 2 3 4 5 6 7 12 11 10 09 08

PREFACE

This *Student Solutions Manual and Study Guide* has been written to accompany the textbook, *College Physics, Eighth Edition,* by Raymond A. Serway, Chris Vuille, and Jerry S. Faughn. The purpose of this ancillary is to provide students with a convenient review of the basic concepts and applications presented in the textbook, together with solutions to selected end-of-chapter problems. This is not an attempt to rewrite the textbook in a condensed fashion. Rather, emphasis is placed upon clarifying typical troublesome points and providing further practice in methods of problem solving.

Every textbook chapter has a matching chapter in this book and each chapter is divided into several parts. Very often, reference is made to specific equations or figures in the textbook. Each feature of this Study Guide has been included to insure that it serves as a useful supplement to the textbook. Most chapters contain the following components:

- **Notes from Selected Chapter Sections:** This is a summary of important concepts, newly defined physical quantities, and rules governing their behavior.

- **Equations and Concepts:** This is a review of the chapter, with emphasis on highlighting important concepts and describing important equations and formalisms.

- **Suggestions, Skills, and Strategies:** This section offers hints and strategies for solving typical problems that the student will often encounter in the course. In some sections, suggestions are made concerning mathematical skills that are necessary in the analysis of problems.

- **Review Checklist:** This is a list of topics and techniques the student should master after reading the chapter and working the assigned problems.

- **Solutions to Selected End-of-Chapter Problems:** Solutions are given for selected odd-numbered problems that were chosen to illustrate important concepts in each textbook chapter.

- **Tables:** A list of selected Physical Constants is printed on the inside front cover; and a table of some Conversion Factors is provided on the inside back cover.

An important note concerning significant figures: The answers to most of the end-of-chapter problems are stated to three significant figures, though calculations are carried out with as many digits as possible. We sincerely hope that this *Student Solutions Manual* *and Study Guide* will be useful to you in reviewing the material presented in the text, and in improving your ability to solve problems and score well on exams. We welcome any comments or suggestions which could help improve the content of this study guide in future editions; and we wish you success in your study.

John R. Gordon
Harrisonburg, VA

Charles Teague
Richmond, KY

Raymond A. Serway
Leesburg, VA

Acknowledgments

We are indebted to everyone who contributed to this *Student Solutions Manual and Study Guide to Accompany College Physics, Eighth Edition.*

We wish to express our sincere appreciation to the staff of Cengage Learning who provided necessary resources and coordinated all phases of this project. Special thanks go to Sylvia Krick (Assistant Freelance Editor), Stefanie Chase (Editorial Assistant), Michelle Cole (Content Project Manager), and Ed Dodd (Development Editor).

Our appreciation goes to our reviewer, Edward Oberhofer, Emeritus Faculty, University of North Carolina at Charlotte. His careful reading of the manuscript and checking the accuracy of the problem solutions contributed in an important way to the quality of the final product. Any errors remaining in the manual are the responsibility of the authors.

It is a pleasure to acknowledge the excellent work of the staff of ICC Macmillan Publishing Solutions for assembling and typesetting this manual and preparing diagrams and page layouts. Their technical skills and attention to detail added much to the appearance and usefulness of this volume.

Finally, we express our appreciation to our families for their inspiration, patience, and encouragement.

Suggestions for Study

We have seen a lot of successful physics students. The question, "How should I study this subject?" has no single answer, but we offer some suggestions that may be useful to you.

1. Work to understand the basic concepts and principles before attempting to solve assigned problems. Carefully read the textbook before attending your lecture on that material. Jot down points that are not clear to you, take careful notes in class, and ask questions. Reduce memorization to a minimum. Memorizing sections of a text or derivations does not necessarily mean you understand the material.

2. After reading a chapter, you should be able to define any new quantities that were introduced and discuss the first principles that were used to derive fundamental equations. A review is provided in each chapter of the Study Guide for this purpose, and the marginal notes in the textbook (or the index) will help you locate these topics. You should be able to correctly associate with each physical quantity the symbol used to represent that quantity (including vector notation, if appropriate) and the SI unit in which the quantity is specified. Furthermore, you should be able to express each important principle or equation in a concise and accurate prose statement. Perhaps the best test of your understanding of the material will be your ability to answer questions and solve problems in the text, or those given on exams.

3. Try to solve plenty of the problems at the end of the chapter. The worked examples in the text will serve as a basis for your study. This Study Guide contains detailed solutions to about twelve of the problems at the end of each chapter. You will be able to check the accuracy of your calculations for any odd-numbered problem, since the answers to these are given at the back of the text.

4. Besides what you might expect to learn about physics concepts, a very valuable skill you can take away from your physics course is the ability to solve complicated problems. The way physicists approach complex situations and break them down into manageable pieces is widely useful. Starting in Section 1.10, the textbook develops a general problem-solving strategy that guides you through the steps. To help you remember the steps of the strategy, they are called *Conceptualize*, *Categorize*, *Analyze*, and *Finalize*.

General Problem-Solving Strategy

Conceptualize

-   The first thing to do when approaching a problem is to *think about* and *understand* the situation. Read the problem several times until you are confident you understand what is being asked. Study carefully any diagrams, graphs, tables, or photographs that accompany the problem. Imagine a movie, running in your mind, of what happens in the problem.

-   If a diagram is not provided, you should almost always make a quick drawing of the situation. Indicate any known values, perhaps in a table or directly on your sketch.

- Now focus on what algebraic or numerical information is given in the problem. In the problem statement, look for key phrases such as "starts from at rest" $(v_i = 0)$, "stops" $(v_f = 0)$, or "freely falls" $(a_y = -g = -9.80 \text{ m/s}^2)$. Key words can help simplify the problem.

- Next focus on the expected result of solving the problem. Exactly what is the question asking? Will the final result be numerical or algebraic? If it is numerical, what units will it have? If it is algebraic, what symbols will appear in it?

- Incorporate information from your own experiences and common sense. What should a reasonable answer look like? What should its order of magnitude be? You wouldn't expect to calculate the speed of an automobile to be 5×10^6 m/s.

Categorize

- Once you have a really good idea of what the problem is about, you need to *simplify* the problem. Remove the details that are not important to the solution. For example, you can often model a moving object as a particle. Key words should tell you whether you can ignore air resistance or friction between a sliding object and a surface.

- Once the problem is simplified, it is important to *categorize* the problem. How does it fit into a framework of ideas that you construct to understand the world? Is it a simple *plug-in problem,* such that numbers can be simply substituted into a definition? If so, the problem is likely to be finished when this substitution is done. If not, you face what we can call an *analysis problem*—the situation must be analyzed more deeply to reach a solution.

- If it is an analysis problem, it needs to be categorized further. Have you seen this type of problem before? Does it fall into the growing list of types of problems that you have solved previously? Being able to classify a problem can make it much easier to lay out a plan to solve it. For example, if your simplification shows that the problem can be treated as a particle moving under constant acceleration and you have already solved such a problem (such as the examples in Section 2.6), the solution to the new problem follows a similar pattern.

Analyze

- Now, you need to analyze the problem and strive for a mathematical solution. Because you have already categorized the problem, it should not be too difficult to select relevant equations that apply to the type of situation in the problem. For example, if your categorization shows that the problem involves a particle moving under constant acceleration, Equations (2.9) to (2.13) are relevant.

- Use algebra (and calculus, if necessary) to solve symbolically for the unknown variable in terms of what is given. Substitute in the appropriate numbers, calculate the result, and round it to the proper number of significant figures.

Finalize

- This final step is the most important part. Examine your numerical answer. Does it have the correct units? Does it meet your expectations from your conceptualization of the problem? What about the algebraic form of the result—before you substituted numerical values? Does it make sense? Try looking at the variables in it to see whether the answer would change in a physically meaningful way if they were drastically increased or decreased or even became zero. Looking at limiting cases to see whether they yield expected values is a very useful way to make sure that you are obtaining reasonable results.

- Think about how this problem compares with others you have done. How was it similar? In what critical ways did it differ? Why was this problem assigned? You should have learned something by doing it. Can you figure out what? Can you use your solution to expand, strengthen, or otherwise improve your framework of ideas? If it is a new category of problem, be sure you understand it so that you can use it as a model for solving future problems in the same category.

When solving complex problems, you may need to identify a series of sub-problems and apply the problem-solving strategy to each. For very simple problems, you probably don't need this whole strategy. But when you are looking at a problem and you don't know what to do next, remember the steps in the strategy and use them as a guide.

Work on problems in this Study Guide yourself and compare your solutions with ours. Your solution does not have to look just like the one presented here. A problem can sometimes be solved in different ways, starting from different principles. If you wonder about the validity of an alternative approach, ask your instructor.

5. We suggest that you use this Study Guide to review the material covered in the text, and as a guide in preparing for exams. You can use the sections Chapter Review, Notes from Selected Chapter Sections, and Equations and Concepts to focus in on any points which require further study. The main purpose of this Study Guide is to improve upon the efficiency and effectiveness of your study hours and your overall understanding of physical concepts. However, it should not be regarded as a substitute for your textbook or for individual study and practice in problem solving.

TABLE OF CONTENTS

15

Electric Forces and Electric Fields

NOTES FROM SELECTED CHAPTER SECTIONS

15.1   Properties of Electric Charges

Electric charge has the following important properties:

- There are two kinds of charges (positive and negative) in nature. Unlike charges attract one another, and like charges repel one another.

- The force between charges varies as the inverse square of the distance between them.

- Charge is always conserved.

- Charge comes in discrete packets (quantized) that are integral multiples of the electronic charge, $e = 1.602\,19 \times 10^{-19}$ C.

A neutral object becomes charged electrically due to the gain or loss of electrons. A negatively charged object has an excess of electrons (relative to protons), and a positively charged object has a deficiency of electrons.

15.2   Insulators and Conductors

Conductors are materials in which electrons move freely under the influence of an electric field; insulators are materials that do not readily transport charge.

15.3   Coulomb's Law

Experiments show that an electric force between two charges is:

- Inversely proportional to the square of the distance between the two charges and is directed along the line joining them.

- Proportional to the product of the magnitudes of the charges.

- Attractive if the charges are of opposite sign and repulsive if the charges have the same sign.

The electric interaction between two charges q_1 and q_2 obeys Newton's third law. The two charges experience forces that are equal in magnitude and opposite in direction. *This is true regardless of the relative magnitudes of the two charges.*

15.4   The Electric Field

An electric field exists at a point if a small test charge placed at that point experiences an electric force. The direction of the electric field is along the direction of the force on a positive test charge. The magnitude of the electric field is given by Equation (15.6). The SI units of the electric field are newtons per coulomb (N/C).

The net electric field at a point (e.g., point P in the figure at right), due to multiple charges, equals the *vector sum* of the electric fields at that point due to the individual charges. *This is called the superposition principle.*

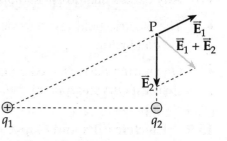

15.5   Electric Field Lines

A convenient aid for visualizing electric field patterns can be obtained by drawing a group of electric field lines. These lines indicate the direction and relative magnitude of the electric field at any point, and are related to the electric field in any region of space in the following manner:

- The electric field vector $\vec{E}$ is tangent to the electric field line at each point.

- The number of electric field lines per unit area through a surface perpendicular to the lines is proportional to the strength of the electric field in that region. Thus, $\vec{E}$ is large when the field lines are close together and small when they are far apart.

The rules for drawing electric field lines for any charge distribution are as follows:

- The lines begin on positive charges and terminate on negative charges, or at infinity in the case of an excess of charge.

- The number of lines drawn leaving a positive charge or terminating on a negative charge is proportional to the magnitude of the charge. In the case of an isolated charge, the field lines are evenly distributed around the charge.

- No two field lines can cross.

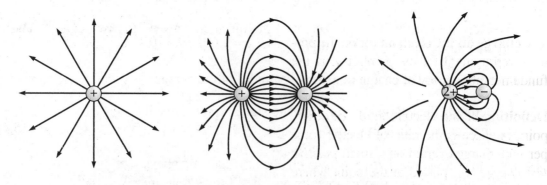

Representative Electric Field Line Patterns

15.6   Conductors in Electrostatic Equilibrium

A conductor in electrostatic equilibrium (no net motion of charge occurs within the conductor) has the following properties:

- The electric field is zero everywhere inside the conductor.

- Any excess charge on an isolated conductor resides entirely on its surface.

- The electric field just outside a charged conductor is perpendicular to the surface of the conductor.

- On an irregularly shaped conductor, the charge per unit area is greatest at locations of sharpest curvature, that is, where the radius of curvature is smallest.

15.9   Electric Flux and Gauss's Law

There is an important general relation between the net electric flux through an imaginary closed surface (called a Gaussian surface) and the charge enclosed by the surface. This is known as Gauss's law and states that the net electric flux through any closed Gaussian surface is equal to the net charge inside the surface divided by ϵ_0. Although Gauss's law is valid for any closed surface, it is most useful for calculating the electric field in situations which have a high degree of symmetry in the charge distribution (for example, uniformly charged spheres, cylinders, lines, or planes).

EQUATIONS AND CONCEPTS

Coulomb's law gives the magnitude of the electrostatic force between two point charges q_1 and q_2, separated by a distance r. This is also true for spherical charge distributions when r is the distance between their centers. In calculations the approximate value $(8.99 \times 10^9 \ \text{N} \cdot \text{m}^2 / \text{C}^2)$ of the Coulomb constant, k_e, may be used. Like sign charges repel and unlike sign charges attract.

$$F = k_e \frac{|q_1||q_2|}{r^2} \qquad (15.1)$$

$$k_e = 8.987 \ 5 \times 10^9 \ \text{N} \cdot \text{m}^2 / \text{C}^2$$

The **charge on the electron** (or on the proton), represented by the symbol e, is the fundamental quantity of charge in nature.

$$e = 1.602 \ 19 \times 10^{-19} \ \text{C}$$

Definition of the electric field, $\vec{\mathbf{E}}$, at any point in space is the ratio of electric force per unit charge exerted on a small *positive test charge*, q_0, placed at the point where the field is to be determined. **The direction of $\vec{\mathbf{E}}$ is along the direction of $\vec{\mathbf{F}}$.** *An electric field exists in the vicinity of charge Q whether or not the test charge is present.*

$$\vec{\mathbf{E}} \equiv \frac{\vec{\mathbf{F}}}{q_0} \qquad (15.4)$$

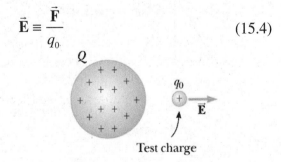

Test charge

The **electric field due to a point charge** has a magnitude given by Equation (15.6). The direction of the electric field is radially outward from a positive point charge and radially inward toward a negative point charge. *The superposition principle holds when the electric field at a point is due to a number of point charges.*

$$E = k_e \frac{|q|}{r^2} \vec{\mathbf{F}} \qquad (15.6)$$

Electric flux through a surface is proportional to the number of electric lines that penetrate the surface. For a plane surface in a uniform field, the flux depends on the angle between the normal to the surface and the direction of the field. Electric flux has SI units of $N \cdot m^2/C$. In the figure, the electric field is parallel to the normal to the plane surface and therefore, in this case, $\theta = 0$ in Equation (15.9).

$$\Phi_E = EA \cos\theta \qquad (15.9)$$

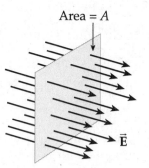

For the orientation shown above the electric flux through the area, A, is EA.

Gauss's law states that the net electric flux, Φ_E, out of a closed surface (Gaussian surface) is equal to the net charge enclosed by the surface divided by the constant ϵ_0. *By convention, for a closed surface, the flux lines passing into the interior of the volume are negative, and those passing out of the interior of the volume are positive.*

$$\Phi_E = \frac{Q_{\text{inside}}}{\epsilon_0} \qquad (15.11)$$

$$Q_{\text{inside}} = Q_{\text{positive}} - Q_{\text{negative}}$$

SUGGESTIONS, SKILLS, AND STRATEGIES

ELECTRIC FORCES AND FIELDS

1. **Units:** When performing calculations using Coulomb's law and the Coulomb constant $k_e = 8.99 \times 10^9 \ N \cdot m^2/C^2$, charge must be in coulombs and distances in meters. If they are given in other units, you must convert them to SI units.

2. **Applying Coulomb's law to point charges:** Remember to use the magnitude of the charges in Equation (15.1). The direction of the force is found by noting that the

forces are repulsive between like charges and attractive between unlike charges. Use the superposition principle properly when dealing with a collection of interacting point charges. When several charges are present, the resultant force on one of them is found by first calculating (magnitude and direction) the force that each of the other charges exerts on it. Then, determine the vector sum of these forces. The magnitude of the force that any charged particle exerts on another is given by Coulomb's law.

3. **Calculating the electric field due to point charges:** The direction of the electric field is the direction of electric force on a positive "test" charge if placed at the point in question. The superposition principle can also be applied to electric fields, which, like electrostatic forces, are also vector quantities. To find the electric field at a given point, first calculate the electric field at the point due to each individual charge. The resultant field at the point is the vector sum of the fields due to the individual charges.

4. **Electric flux and Gauss's law:** When using Gauss's law to calculate the electric field in the vicinity of a charge distribution, you must anticipate the expected symmetry of the electric field by considering the shape of the charge distribution. Choose a Gaussian surface which has a shape matching the symmetry of the field (spherical surface for a point charge, cylindrical surface for a line of charge, etc.). The Gaussian surface must be closed and include the point at which the field is to be calculated.

An important factor in your choice of a shape should be the ease with which you can calculate the flux through the closed surface. This will be the case when the surface can be divided into regions so that over a particular region either (1) the flux is zero, or (2) the electric field is constant. Consider, for example, a line of charge along the axis of a closed cylindrical surface.

REVIEW CHECKLIST

You should be able to:

- Use Coulomb's law to determine the net electrostatic force (magnitude and direction) on a point electric charge due to a known distribution of a finite number of point charges. (Section 15.3)

- Calculate the electric field $\vec{E}$ (magnitude and direction) at a specified location in the vicinity of a group of point charges. (Section 15.4)

- Describe the pattern of electric field lines associated with charge distributions such as (i) single point charges, (ii) pairs of charges with like or unlike sign and equal or unequal magnitudes, (iii) charged metallic sphere, etc. (Section 15.5)

- State and justify the conditions for the charge distribution on a conductor in electrostatic equilibrium. (Section 15.6)

- Calculate the electric flux through a surface and use Gauss's law to determine the electric field due to symmetric charge distributions (point, sphere, line, cylinder, plane). (Section 15.9)

SOLUTIONS TO SELECTED END-OF-CHAPTER PROBLEMS

5. The nucleus of ^{8}Be, which consists of 4 protons and 4 neutrons, is very unstable and spontaneously breaks into two alpha particles (helium nuclei, each consisting of 2 protons and 2 neutrons). (a) What is the force between the two alpha particles when they are 5.00×10^{-15} m apart, and (b) what is the initial magnitude of the acceleration of the alpha particles due to this force? Note that the mass of an alpha particle is 4.002 6 u.

Solution

(a) Treat the two alpha particles, resulting from the beryllium decay, as point charges with

$$q_1 = q_2 = +2e = 2(1.60 \times 10^{-19} \text{ C}) = 3.20 \times 10^{-19} \text{ C}$$

When separated by a distance $r = 5.00 \times 10^{-15}$ m, these positive charges repel each other with a force whose magnitude is given by Coulomb's law:

$$F = k_e \frac{|q_1||q_2|}{r^2} = (8.99 \times 10^9 \text{ N} \cdot \text{m}^2/\text{C}^2) \frac{(3.20 \times 10^{-19} \text{ C})^2}{(5.00 \times 10^{-15} \text{ m})^2} = 36.8 \text{ N} \qquad \lozenge$$

(b) The mass of an alpha particle is $\qquad\qquad m = 4.002\ 6 \text{ u}$

where the unified mass unit is given by $\qquad\qquad 1\text{u} = 1.66 \times 10^{-27} \text{ kg}$

Newton's second law then gives the magnitude of the acceleration of each alpha particle as

$$a = \frac{F}{m} = \frac{36.8 \text{ N}}{4.002\ 6(1.66 \times 10^{-27} \text{ kg})} = 5.54 \times 10^{27} \text{ m/s}^2 \qquad \lozenge$$

---

9. Two small identical conducting spheres are placed with their centers 0.30 m apart. One is given a charge of 12×10^{-9} C, the other a charge of -18×10^{-9} C. (a) Find the electrostatic force exerted on one sphere by the other. (b) The spheres are connected by a conducting wire. Find the electrostatic force between the two after equilibrium is reached.

Solution

(a) Since the charged spheres are small, they may be considered as point charges located at their centers. Coulomb's law then gives the magnitude of the force one sphere exerts on the other as

$$F = k_e \frac{|q_1||q_2|}{r^2} = (8.99 \times 10^9 \text{ N} \cdot \text{m}^2/\text{C}^2) \frac{(12 \times 10^{-9} \text{ C})(18 \times 10^{-9} \text{ C})}{(0.30 \text{ m})^2} = 2.2 \times 10^{-5} \text{ N} \qquad \lozenge$$

This is an attractive force since the two spheres possess unlike charges. $\qquad \lozenge$

(b)   When a conducting wire connects the two spheres, charge will flow from one sphere to the other until equilibrium is established. Since the two spheres are identical, the total excess charge on the two spheres will be divided equally between the two spheres when equilibrium occurs. Therefore, each sphere will then have a net charge of

$$q_f = \frac{q_1 + q_2}{2} = \frac{12 \times 10^{-9} \text{ C} - 18 \times 10^{-9} \text{ C}}{2} = -3.0 \times 10^{-9} \text{ C}$$

After equilibrium is reached, the two spheres will repel each other with forces of magnitude

$$F' = k_e \frac{|q_f||q_f|}{r^2} = (8.99 \times 10^9 \text{ N} \cdot \text{m}^2/\text{C}^2) \frac{(3.0 \times 10^{-9} \text{ C})(3.0 \times 10^{-9} \text{ C})}{(0.30 \text{ m})^2} = 9.0 \times 10^{-7} \text{ N} \quad \Diamond$$

---

14. A charge of −3.00 nC and a charge of −5.80 nC are separated by a distance of 50.0 cm. Find the position at which a third charge of +7.50 nC can be placed so that the net electrostatic force on it is zero.

Solution

The positive third charge will experience a force of attraction toward each of the two negative charges. If the vector sum of these two forces is to be zero the two forces must have equal magnitudes and be in opposite directions. These attractive forces will be in opposite directions only when the positive charge is located between the two negative charges on the line connecting them as shown in the sketch above.

The position the positive charge should have along this connecting line is determined by the requirement that the two attractive forces have equal magnitudes. Thus, we require that $F_2 = F_1$ and look for a solution having $0 < x < 50.0$ cm, where x is the distance of the positive charge q_3 from the charge $q_1 = -3.00$ nC as shown in the sketch. This gives

$$k_e \frac{|q_2||q_3|}{(50.0 \text{ cm} - x)^2} = k_e \frac{|q_1||q_3|}{x^2} \quad \text{which reduces to} \quad x^2 = \left|\frac{q_1}{q_2}\right|(50.0 \text{ cm} - x)^2$$

or $x = \pm\sqrt{\dfrac{3.00 \text{ nC}}{5.80 \text{ nC}}}(50.0 \text{ cm} - x) = \pm 0.719(50.0 \text{ cm} - x)$

Using the upper sign gives $x = 36.0 \text{ cm} - 0.719x$

or $x = \dfrac{+36.0 \text{ cm}}{1.719} = +20.9$ cm which is the desired location with $0 < x < 50.0$ cm. $\Diamond$

Using the lower sign leads to $x = -36.0 \text{ cm} + 0.719x$ or $x = \dfrac{-36.0 \text{ cm}}{0.281} = -128$ cm

At this location, 128 cm to the left of the −3.00 nC charge, the two forces would indeed have equal magnitudes. However, they would have the same direction (toward the right) and could not add to zero. Thus, this solution must be rejected.

---

23. A proton accelerates from rest in a uniform electric field of 640 N/C. At some later time, its speed is 1.20×10^6 m/s. (a) Find the magnitude of the acceleration of the proton. (b) How long does it take the proton to reach this speed? (c) How far has it moved in that interval? (d) What is its kinetic energy at the later time?

Solution

(a) The force exerted on the proton by the electric field has magnitude $F = qE$. The charge and mass of the proton are:

$$q = +e = 1.60 \times 10^{-19} \text{ C}$$

and $m_p = 1.673 \times 10^{-27}$ kg

Newton's second law then gives the acceleration as

$$a = \frac{F}{m_p} = \frac{eE}{m_p} = \frac{(1.60 \times 10^{-19} \text{ C})(640 \text{ N/C})}{1.673 \times 10^{-27} \text{ kg}} = 6.12 \times 10^{10} \text{ m/s}^2 \qquad \Diamond$$

(b) The force, and hence the acceleration, is constant in a uniform electric field. Therefore, the time required to reach the final speed is

$$t = \frac{v_f - v_i}{a} = \frac{1.20 \times 10^6 \text{ m/s} - 0}{6.12 \times 10^{10} \text{ m/s}^2} = 1.96 \times 10^{-5} \text{ s} = 19.6 \text{ } \mu s \qquad \Diamond$$

(c) The uniformly accelerated motion equation $v_f^2 = v_i^2 + 2a(\Delta x)$ gives the displacement of the proton during this time as

$$\Delta x = \frac{v_f^2 - v_i^2}{2a} = \frac{(1.20 \times 10^6 \text{ m/s})^2 - 0}{2(6.12 \times 10^{10} \text{ m/s}^2)} = 11.8 \text{ m} \qquad \Diamond$$

(d) The kinetic energy of the proton at the end of this time interval is

$$KE_f = \frac{1}{2} m_p v_f^2 = \frac{1}{2}(1.673 \times 10^{-27} \text{ kg})(1.20 \times 10^6 \text{ m/s})^2 = 1.20 \times 10^{-15} \text{ J} \qquad \Diamond$$

This result can also be obtained from the work-energy theorem $W_{net} = KE_f - KE_i$, where $W_{net} = F(\Delta x) = eE(\Delta x)$, as follows:

$$KE_f = W_{net} + KE_i = eE(\Delta x) + 0$$

or $KE_f = (1.60 \times 10^{-19} \text{ C})(640 \text{ N/C})(11.8 \text{ m}) = 1.21 \times 10^{-15} \text{ J} \qquad \Diamond$

The difference in the two values found for the final kinetic energy is due to the rounding error in the answer to part (c).

---

27. In Figure P15.27, determine the point (other than infinity) at which the total electric field is zero.

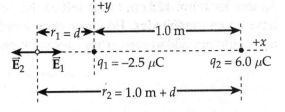

Solution

At each point in space, the electric field generated by the charge distribution shown in Figure P15.27 is the vector sum of two contributions, one due to each of the charges q_1 and q_2. In order for this field to be zero, the two contributions must have equal magnitudes and opposite directions.

The contribution $\vec{E}_1$ due to the negative charge q_1 will be directed toward q_1, while the contribution $\vec{E}_2$ due to the positive charge q_2 is directed away from that charge. Only on the straight line passing through both charges (the x-axis in the above sketch) will these contributions lie along the same line. In the region of this line between the two charges, the two contributions are in the same direction and cannot add to zero. Thus, the point of interest must lie on the x-axis, either to the left of q_1 or to the right of q_2. If the two contributions to the resultant field are to have equal magnitudes, the point of interest must be located nearer to the smaller charge q_1. Thus, it must be located on the portion of the x-axis to the left of q_1 as shown in the sketch.

Using the notation given in the sketch, and requiring that $E_1 = E_2$, or $k_e \dfrac{|q_1|}{r_1^2} = k_e \dfrac{|q_2|}{r_2^2}$, we obtain

$$k_e \frac{2.5 \ \mu C}{d^2} = k_e \frac{6.0 \ \mu C}{\left(1.0 \text{ m} + d\right)^2} \quad \text{or} \quad 1.0 \text{ m} + d = \pm d \sqrt{\frac{6.0}{2.5}} = \pm 1.55 d$$

Using the upper sign gives $1.0 \text{ m} = 0.55d$ and $d = \dfrac{+1.0 \text{ m}}{0.55} = +1.8 \text{ m}$

so the point where the field is zero is on the x-axis, 1.8 m to the left of q_1.   ◊

Note that using the lower sign would have given $d = \dfrac{-1.0 \text{ m}}{2.55} = -0.39$ m, which is located between the two charges (since $d > 0$ is to the left of q_1 according to the notation adopted in the sketch). Here E_1 and E_2 have the same magnitudes, but they also the same direction and cannot add to zero. Thus, this is not an acceptable solution.

33. Two point charges are a small distance apart. (a) Sketch the electric field lines for the two if one has a charge four times that of the other and both charges are positive. (b) Repeat for the case in which both charges are negative.

Solution

In sketching the electric field patterns for given charge distributions, there are several points one should keep in mind.

Some of these are:

(1)   When point charges are isolated in space, the field lines exhibit radial symmetry. That is, they are uniformly spaced around the charge and either radially outward away from the charge or radially inward toward the charge. When in the presence of other charges, the field lines will still be nearly radial at points very close to the charge but will deviate from radial farther out.

(2)   Field lines originate on positive charges and terminate on negative charges.

(3)   The number of field lines one should draw leaving a positive charge or approaching a negative charge is proportional to the charge. Therefore, the number of lines leaving (or approaching) a charge of magnitude $4q$ should be four times the number leaving (or approaching) a charge of magnitude q.

(4)   Field lines never cross each other.

   (a)   In the drawing below for two point charges of magnitudes q and $4q$, located near each other, not all of the lines leaving the charge $4q$ are shown for the sake of clarity in the drawing.   ◊

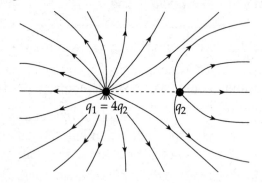

   (b)   If the charges q_1 and q_2 were negative instead of positive, the pattern of field lines would be as shown above with the exception that the arrows would point in the opposite direction along the lines (that is, in toward the negative charges).   ◊

41. An electric field of intensity **3.50 kN/C** is applied along the *x*-axis. Calculate the electric flux through a rectangular plane 0.350 m wide and 0.700 m long if (a) the plane is parallel to the *yz*-plane; (b) the plane is parallel to the *xy*-plane; and (c) the plane contains the *y*-axis, and its normal makes an angle of 40.0° with the *x*-axis.

Solution

The planar section of interest has area
$$A = length \times width = (0.700 \text{ m}) \times (0.350 \text{ m}) = 0.245 \text{ m}^2$$

When this area is in a uniform electric field of magnitude *E*, the flux through it will be

$$\Phi_E = EA\cos\theta$$

where θ is the angle that the perpendicular or "normal" to the area makes with the direction of the electric field (the positive *x*-axis in this case).

(a) If the area is parallel to the *yz*-plane, then its perpendicular is parallel to the *x*-axis, and hence, to the field. In this case, $\theta = 0°$, and the flux is

$$\Phi_E = (3.50 \times 10^3 \text{ N/C})(0.245 \text{ m}^2)\cos 0° = 858 \text{ N} \cdot \text{m}^2/\text{C} \qquad \Diamond$$

(b) When the area is parallel to the *xy*-plane, its normal is parallel to the *z*-axis, and hence makes an angle of 90.0° with the direction of the field. The flux is then

$$\Phi_E = (3.50 \times 10^3 \text{ N/C})(0.245 \text{ m}^2)\cos 90.0° = 0 \qquad \Diamond$$

(c) If the plane of the area contains the *y*-axis and its normal makes an angle of 40.0° with the *x*-axis, then $\theta = 40.0°$ and the flux is

$$\Phi_E = (3.50 \times 10^3 \text{ N/C})(0.245 \text{ m}^2)\cos 40.0° = 657 \text{ N} \cdot \text{m}^2/\text{C} \qquad \Diamond$$

---

45. A point charge *q* is located at the center of a spherical shell of radius *a* that has a charge –*q* uniformly distributed on its surface. Find the electric field (a) for all points outside the spherical shell and (b) for a point inside the shell a distance *r* from the center.

Solution

With the positive point change at the center of the spherical shell, and the negative charge uniformly distributed on the surface of the shell, the charge distribution has complete spherical symmetry. Thus, anywhere that an electric field due to this distribution exists, that field will be in the radial direction, and its magnitude will depend only on the distance *r* from the center of the spherical shell.

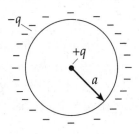

To take advantage of the symmetry known to exist in the field, we shall employ a spherical Gaussian surface that is concentric with the spherical shell. The flux through this surface is then

$$\Phi_E = EA\cos\theta = E\left(4\pi r^2\right)\cos 0° = 4\pi r^2 E$$

where r is the radius of the Gaussian surface, and E is the magnitude of the electric field at all points on this surface.

Gauss's law also tells us that the flux through our Gaussian surface is equal to the net charge enclosed by the surface divided by the permittivity of free space, $\in_0$. Thus, we have

$$\Phi_E = 4\pi r^2 E = Q_{inside}/\in_0 \qquad \text{or} \qquad E = Q_{inside}/4\pi\in_0 r^2$$

(a) If our Gaussian surface is outside the spherical shell, $r > a$, then the net charge inside the closed surface is $Q_{inside} = +q - q = 0$. Thus, the electric field at all points outside the spherical shell has zero magnitude.   ◊

(b) If the radius of our Gaussian surface is less than that of the shell, $0 < r < a$, then $Q_{inside} = +q$, and the electric field at points inside the shell is

$$\vec{E} = q/4\pi\in_0 r^2 \quad \text{radially outward (away from the positive charge)} \qquad ◊$$

======

48. A very large nonconducting plate lying in the *xy*-plane carries a charge per unit area of σ. A second such plate located at $z = 2.00$ cm and oriented parallel to the *xy*-plane carries a charge per unit area of -2σ. Find the electric field (a) for $z < 0$, (b) $0 < z < 2.00$ cm, and (c) $z > 2.00$ cm.

Solution

Before solving this problem, the reader should carefully review Example 15.8 in the textbook. There, Gauss's law is used to determine the electric field produced by a nonconducting plane sheet of charge. The results of that analysis are that the field has a constant magnitude given by $E = \left|\sigma_{sheet}\right|/2\in_0$ (where σ_{sheet} is the charge per unit area on the sheet), is everywhere perpendicular to the sheet, is directed away from the sheet if $\sigma_{sheet} > 0$, and is directed toward the sheet if $\sigma_{sheet} < 0$.

In this problem, we have two plane sheets of charge, both parallel to the *xy*-plane and separated by a distance of 2.00 cm. The upper sheet has charge density $\sigma_{sheet} = -2\sigma$, while the lower sheet has $\sigma_{sheet} = +\sigma$. At any location, the resultant electric field is the vector sum of the fields due to the two individual sheets of charge. Taking upward as the positive *z*-direction, the fields due to each of the sheets in the three regions of interest are:

	Lower sheet (at $z = 0$)	**Upper sheet (at $z = 2.00$ cm)**
Region	Electric Field	Electric Field
$z < 0$	$E_z = -\dfrac{\lvert +\sigma \rvert}{2\,\epsilon_0} = -\dfrac{\sigma}{2\,\epsilon_0}$	$E_z = +\dfrac{\lvert -2\sigma \rvert}{2\,\epsilon_0} = +\dfrac{\sigma}{\epsilon_0}$
$0 > z > 2.00$ cm	$E_z = +\dfrac{\lvert +\sigma \rvert}{2\,\epsilon_0} = +\dfrac{\sigma}{2\,\epsilon_0}$	$E_z = +\dfrac{\lvert -2\sigma \rvert}{2\,\epsilon_0} = +\dfrac{\sigma}{\epsilon_0}$
$z > 2.00$ cm	$E_z = +\dfrac{\lvert +\sigma \rvert}{2\,\epsilon_0} = +\dfrac{\sigma}{2\,\epsilon_0}$	$E_z = -\dfrac{\lvert -2\sigma \rvert}{2\,\epsilon_0} = -\dfrac{\sigma}{\epsilon_0}$

The resultant electric field in each of the regions of interest is then:

(a) For $z < 0$: $E_z = E_{z,\text{lower}} + E_{z,\text{upper}} = -\dfrac{\sigma}{2\,\epsilon_0} + \dfrac{\sigma}{\epsilon_0} = +\dfrac{\sigma}{2\,\epsilon_0}$ ◊

(b) For $0 > z > 2.00$ cm: $E_z = E_{z,\text{lower}} + E_{z,\text{upper}} = +\dfrac{\sigma}{2\,\epsilon_0} + \dfrac{\sigma}{\epsilon_0} = +\dfrac{3\sigma}{2\,\epsilon_0}$ ◊

(c) For $z > 2.00$ cm: $E_z = E_{z,\text{lower}} + E_{z,\text{upper}} = +\dfrac{\sigma}{2\,\epsilon_0} - \dfrac{\sigma}{\epsilon_0} = -\dfrac{\sigma}{2\,\epsilon_0}$ ◊

---

53. (a) Two identical point charges $+q$ are located on the y-axis at $y = +a$ and $y = -a$. What is the electric field along the x-axis at $x = b$? (b) A circular ring of charge of radius a has a total positive charge Q distributed uniformly around it. The ring is in the $x = 0$ plane with its center at the origin. What is the electric field along the x-axis at $x = b$ due to the ring of charge? [Hint: Consider the charge Q to consist of many pairs of identical point charges positioned at the ends of diameters of the ring.]

Solution

(a) Any point on the x-axis is equidistant from the two identical point charges. Thus, the contributions by the two charges to the resultant field at this point have equal magnitudes given by

$$E_1 = E_2 = \frac{k_e q}{r^2}$$

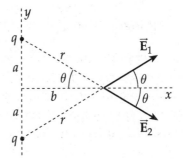

The components of the resultant field are

$$E_y = E_{1y} - E_{2y}: \qquad E_y = \left(\frac{k_e q}{r^2}\right)\sin\theta - \left(\frac{k_e q}{r^2}\right)\sin\theta = 0$$

and $E_x = E_{1x} + E_{2x}$: $\qquad E_x = \left(\dfrac{k_e q}{r^2}\right)\cos\theta + \left(\dfrac{k_e q}{r^2}\right)\cos\theta = \left[\dfrac{k_e(2q)}{r^2}\right]\cos\theta$

Since $\qquad\qquad\qquad\qquad r = \sqrt{a^2 + b^2}$ and $\cos\theta = b/r$

the net x-component becomes $\quad E_x = \dfrac{k_e(2q)b}{r^3} = \dfrac{k_e b(2q)}{\left(a^2 + b^2\right)^{3/2}}$

The resultant electric field on the axis at $x = b$ is:

$$\vec{\mathbf{E}}_R = \frac{k_e b(2q)}{\left(a^2 + b^2\right)^{3/2}} \quad \text{in the } + x \text{ direction} \qquad\qquad \Diamond$$

(b) Note that the result of part (a) may be written as $\vec{\mathbf{E}}_R = \dfrac{k_e b(Q)}{\left(a^2 + b^2\right)^{3/2}}$ in the $+x$ direction

where $Q = 2q$ is the total charge in the charge distribution generating the field.

In the case of a uniformly charged circular ring, consider the ring to consist of many pairs of identical charges, located on opposite ends of a diameter of the ring and uniformly spaced around the ring. Each pair of charges has a total charge Q_i. At a point on the axis of the ring, this pair of charges generates an electric field contribution that is parallel to the axis and has magnitude

$$E_i = \frac{k_e b Q_i}{\left(a^2 + b^2\right)^{3/2}}$$

The resultant electric field of the ring is the summation of the contributions by all such pairs of charges, or

$$E_R = \Sigma E_i = \left[\frac{k_e b}{\left(a^2 + b^2\right)^{3/2}}\right]\Sigma Q_i = \frac{k_e b Q}{\left(a^2 + b^2\right)^{3/2}}$$

where $Q = \Sigma Q_i$ is the total charge on the ring.

Thus, the electric field at a point on the axis of the ring (radius a), and distance b from the plane of the ring, is directed parallel to the axis and away from the ring. Its magnitude is

$$E_R = \frac{k_e b Q}{\left(a^2 + b^2\right)^{3/2}} \qquad\qquad\qquad \Diamond$$

59. Two hard rubber spheres of mass 15 g are rubbed vigorously with fur on a dry day and are then suspended from a rod with two insulating strings of length 5.0 cm. They are observed to hang at equilibrium as shown in Figure P15.59, each at an angle of 10° with the vertical. Estimate the amount of charge that is found on each sphere. (Problem 59 is courtesy of E.F. Redish. For more problems of this type, visit http://www.physics.umd.edu/perg/.)

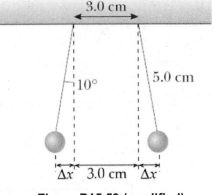

Figure P15.59 (modified)

Solution

Since each sphere has been treated the same as the other, we assume the spheres have been given identical charges (i.e., that $q_1 = q_2 = q$). Each sphere then exerts a repulsive force of magnitude $F_e = k_e \dfrac{|q_1||q_2|}{r^2} = k_e \dfrac{|q|^2}{r^2}$ on the other. Consider the free-body diagram of one of the spheres given at the right. Here, T is the tension in the string supporting the sphere. With the spheres in equilibrium in the positions shown, Newton's second law gives

$$\Sigma F_y = T\cos 10° - mg = 0 \quad \text{or} \quad T = \frac{mg}{\cos 10°}$$

and $\quad \Sigma F_x = T\sin 10° - F_e = 0 \quad$ or $\quad F_e = T\sin 10° = \left(\frac{mg}{\cos 10°}\right)\sin 10° = (mg)\tan 10°$

Thus, $\quad k_e \dfrac{|q|^2}{r^2} = (mg)\tan 10° \quad$ and $\quad |q| = r\sqrt{\dfrac{(mg)\tan 10°}{k_e}}$

where $r = 3.0 \text{ cm} + 2(\Delta x) = 3.0 \text{ cm} + 2\left[(5.0 \text{ cm})\sin 10°\right] = 4.7 \text{ cm}$ is the equilibrium distance between the centers of the two spheres. Our estimate of the charge on each sphere is then

$$|q| = (0.047 \text{ m})\sqrt{\frac{(15 \times 10^{-3} \text{ kg})(9.80 \text{ m/s}^2)\tan 10°}{8.99 \times 10^9 \text{ N} \cdot \text{m}^2/\text{C}^2}} = 8.0 \times 10^{-8} \text{ C} \quad \text{or} \quad |q| \sim 10^{-7} \text{ C} \qquad \Diamond$$

63. Each of the electrons in a particle beam has a kinetic energy of 1.60×10^{-17} J. (a) What is the magnitude of the uniform electric field (pointing in the direction of the electrons' movement) that will stop these electrons in a distance of 10.0 cm? (b) How long will it take to stop the electrons? (c) After the electrons stop, what will they do? Explain.

Solution

(a)   When a negatively charged particle, such as an electron, is in an electric field, it experiences a force of magnitude $F_e = |q|E$ in the direction opposite to that of the field. Thus, the electrons in this particle beam experience a retarding force of magnitude $F_e = eE$. When an electron moves a distance $\Delta x = 0.100$ m against this retarding force, the work done on it by the field is $W = F_e(\Delta x)\cos 180° = -eE(\Delta x)$. If the electrons are brought to rest $(KE_f = 0)$ in this distance, the work-energy theorem gives $W = \Delta KE$, or

$$-eE(\Delta x) = 0 - KE_i \qquad \text{and} \qquad E = \frac{KE_i}{e(\Delta x)}$$

The required magnitude of the field is then

$$E = \frac{1.60 \times 10^{-17} \text{ J}}{(1.60 \times 10^{-19} \text{ C})(0.100 \text{ m})} = 1.00 \times 10^3 \text{ N/C} \qquad \lozenge$$

(b)   The acceleration of the electrons is $a_x = -F_e/m_e = -eE/m_e$. The time required for them to come to rest is

$$t = \frac{v_f - v_i}{a_x} = \frac{0 - v_i}{-eE/m_e} = \frac{m_e v_i}{eE} = \frac{\sqrt{2m_e(KE_i)}}{eE}$$

or $t = \dfrac{\sqrt{2(9.11 \times 10^{-31} \text{ kg})(1.60 \times 10^{-17} \text{ J})}}{(1.60 \times 10^{-19} \text{ C})(1.00 \times 10^3 \text{ N/C})} = 3.37 \times 10^{-8}$ s $= 33.7$ ns $\lozenge$

(c)   After bringing the electron to rest, the electric force continues to act on it, causing it to accelerate in the direction opposite to the field (hence, opposite to the direction of the electron's original motion) at a rate of

$$|a_x| = \frac{eE}{m_e} = -\frac{(1.60 \times 10^{-19} \text{ C})(1.00 \times 10^3 \text{ N/C})}{9.11 \times 10^{-31} \text{ kg}} = 1.76 \times 10^{14} \text{ m/s}^2 \qquad \lozenge$$

16

Electrical Energy and Capacitance

NOTES FROM SELECTED CHAPTER SECTIONS

16.1   Potential Difference and Electric Potential

The electrostatic force is conservative; it is possible to define an electric potential energy function associated with this force. *The potential difference between two points is proportional to the change in potential energy of a charge as it moves between the two points.*

Electrical potential difference (a scalar quantity) is the negative of the work done by the electric field (the conservative force) in moving a charge from point *A* to point *B* divided by the magnitude of the charge. The SI unit of potential is the joule per coulomb or volt (V).

A positive charge gains electrical potential energy when it is moved opposite to the direction of the electric field. When a positive charge is placed in an electric field, it moves in the direction of the field, from a point of high potential to a point of lower potential.

16. 2   Electric Potential and Potential Energy Due to Point Charges

In electric circuits, a point of zero potential is often defined by grounding (connecting to Earth) some point in the circuit. In the case of a point charge, the point of zero potential is taken to be at an infinite distance from the charge. The electric potential at a given point in space due to a point charge *q* depends only on the value of the charge and the distance, *r*, from the charge to the specified point in space. An electric potential can exist at a point in space whether or not a test charge exists at that point.

When using the superposition principle to determine the value of the electric potential at a point due to several point charges, the **algebraic sum** of the individual electric potentials must be used. *Be careful not to confuse electric potential difference with electric potential energy.*

Electric potential is a scalar property of the region surrounding an electric charge. It does not depend on the presence of a test charge in the field.

Electric potential energy is a characteristic of a charge-field system. It is due to the interaction of the field and a charge located within the field.

In the figure to the right an electric field (due to the presence of $+Q$) is present in the region surrounding the charge. This field is due to a point charge and has the properties described in Chapter 15.

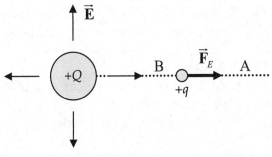

Refer to the figure as you consider the following statements about electric potential difference and electric potential energy.

- Electric field lines, four of which are shown, are directed along outward radials from the charge Q. Points "A" and "B" are shown along one of the field lines.

- If a positive test charge q is placed at some point between A and B, the electric field will exert a force $\vec{F}_E = q\vec{E}$ on q directed along the field line and away from Q.

- If the test charge q is moved from point A to point B (where the potential is greater), the electric field will do work, $W = -q\Delta V = -q(V_B - V_A)$, and the electric potential energy of the two-charge system (Q and q) increases by an amount equal to $q\Delta V$. *The electric potential energy of a positive charge increases when the charge moves to a point closer to another isolated positive point charge (or along a direction opposite the direction of the electric field).*

- *Since the electrostatic force is a conservative force, the work done by the electric field as the test charge q moves from point A to point B will be the same regardless of the actual path taken.*

- You should consider the result of changing either or both Q and q from a positive to a negative charge.

16.3   Potentials and Charged Conductors

16.4   Equipotential Surfaces

A surface on which every point has the same value of electric potential is an **equipotential surface.** The electric field at every point on an equipotential surface is perpendicular to the surface because a parallel component of $\vec{E}$ would imply work would be required to move a charge between two points on the surface.

No work is required to move a charge between two points that are at the same potential. That is, $W = 0$ when $V_B = V_A$.

The electric potential is a constant everywhere on the surface of a charged conductor in electrostatic equilibrium.

The electric potential is constant everywhere inside a conductor, and equal to its value at the surface. This is true because the electric field is zero inside a conductor and no work is required to move a charge between any two points inside the conductor.

The electron volt (eV) is defined as the kinetic energy that an electron (or proton) gains (or loses) when accelerated through a potential difference of 1 V.

16.6   Capacitance

A capacitor is a device consisting of a pair of conductors (plates) separated by insulating material. A charged capacitor acts as a storehouse of charge and energy in an electric field. The capacitance of a capacitor depends on the following physical characteristics of the device: size, shape, plate separation, and the nature of the dielectric medium filling the region between the plates.

The capacitance, *C*, of a capacitor is defined as the ratio of the magnitude of the charge on either conductor to the magnitude of the potential difference between the conductors. Capacitance has SI units of coulombs per volt, called **farads** (F). The farad is a very large unit of capacitance. In practice, most capacitors have capacitances ranging from picofarads to microfarads.

16.8 Combinations of Capacitors

Two or more capacitors can be connected in a circuit in several possible combinations. For example, three capacitors (each assumed to have the same value of capacitance, *C*) can be combined as illustrated in the figure at right to achieve four different values of equivalent capacitance.

(a) all in parallel

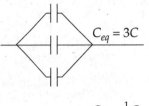

$C_{eq} = 3C$

(b) all in series

$C_{eq} = \frac{1}{3}C$

(c) 2 in series, in parallel with a third

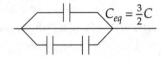

$C_{eq} = \frac{3}{2}C$

(d) 2 in parallel, in series with a third

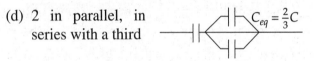

$C_{eq} = \frac{2}{3}C$

Can you determine how many different values of equivalent capacitance could be achieved by combining the three capacitors if they each had a different capacitance, C_1, C_2, and C_3?

Note in the figure above that when a group of capacitors is connected in series, they are arranged "end-to-end," and only adjacent capacitors have a circuit point in common. When connected in parallel, each capacitor in the group has two circuit points in common with each of the others.

For capacitors connected in series:

- The reciprocal of the equivalent capacitance equals the sum of the reciprocals of the individual capacitances. See Equation (16.15).

- The potential difference across the series group equals the sum of the potential differences across the individual capacitors.

- Each capacitor of a group in series has the same charge (even if their capacitances are not the same), and this is also the value of the total charge on the series group.

For capacitors connected in parallel:

- The equivalent capacitance equals the sum of the individual capacitances. See Equation (16.12).

- The potential difference across each capacitor has the same value.

- The total charge on the parallel group equals the sum of the charges on the individual capacitors.

16.9   Energy Stored in a Charged Capacitor

The energy stored in a given charged capacitor is proportional to the square of the potential difference between the two plates.

Each capacitor has a limiting voltage that depends on the capacitor's physical characteristics. When the potential difference between the plates of a capacitor exceeds that limiting voltage, a discharge will occur through the insulating material between the two plates. This is known as electrical breakdown. Therefore, the maximum energy which can be stored in a capacitor is limited by the breakdown voltage.

16.10   Capacitors with Dielectrics

A dielectric is an insulating material, such as rubber, glass, waxed paper, or air. When a dielectric is inserted between the plates of a capacitor, the capacitance increases. If the dielectric completely fills the space between the plates, the capacitance is multiplied by a dimensionless factor κ, called the **dielectric constant.** The dielectric constant is a property of the dielectric material and has a value of 1.000 for a vacuum.

The smallest plate separation for a capacitor is limited by the electric discharge that can occur through the dielectric material separating the plates. For any given plate separation, there is a maximum electric field that can be produced in the dielectric before it breaks down and begins to conduct. This maximum electric field is called the **dielectric strength** and has SI units of V/m.

EQUATIONS AND CONCEPTS

The **change in electric potential energy** of an electric charge in moving between two points in an electric field is equal to the negative of the work done by the electric force.

$$\Delta PE = -W_{AB} = -\left(qE_x\right)\Delta x \qquad (16.1)$$

In a **uniform electric field** the change in electric potential energy can be expressed in terms of the charge, the magnitude of the field, and the distance moved parallel to the direction of the field.

$$\Delta PE = -qEd$$
(in a uniform electric field)

The **electric potential difference** between points A and B is defined as the change in electric potential energy per unit charge as a charge is moved from point A to point B. The SI unit of electric potential is the volt, V.

$$\Delta V \equiv V_B - V_A = \frac{\Delta PE}{q} \qquad (16.2)$$

$$1\ V \equiv 1\ J/C$$

The **potential difference in a uniform field** (constant in magnitude and direction) depends only on the displacement Δx in a direction parallel to the field. *Electric field lines always point along the direction of decreasing potential.*

$$\Delta V = -E_x \Delta x \tag{16.3}$$

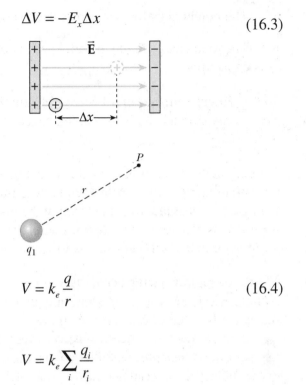

The **electric potential due to a point charge** is inversely proportional to the distance from the charge. *This equation assumes that the potential at infinity is zero. Remember that electric potential is a scalar and the correct algebraic sign for the charge, q, must be used in Equation (16.4).*

$$V = k_e \frac{q}{r} \tag{16.4}$$

The **total electric potential due to several point charges** is the algebraic sum of the potentials due to the individual charges.

$$V = k_e \sum_i \frac{q_i}{r_i}$$

The **potential energy of a pair of charges** separated by a distance, r, represents the minimum work required to assemble the charges from an infinite separation. *The potential energy of the two charges is positive if the two charges have the same sign; and it is negative if the two charges are of opposite sign.*

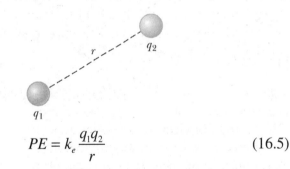

$$PE = k_e \frac{q_1 q_2}{r} \tag{16.5}$$

The **work required to move a charge** q from point A to point B in an electric field depends on the magnitude and sign of the charge. *Zero work is required to move a charge between two points that are at the same potential.*

$$W = -q(V_B - V_A) \tag{16.6}$$

Important characteristics of a charged conductor in electrostatic equilibrium:

- All excess charge resides on the surface.

- $\vec{E} = 0$ at all interior points.

- $\vec{E}$ is perpendicular to the surface on the outside.

- Electric potential is constant in the interior and equal to its value on the surface.

The **capacitance** (C) **of a capacitor** is defined as the ratio of the magnitude of the charge on either plate (conductor) to the potential difference between the plates (C is always positive). The SI unit of capacitance is the farad (F).

1 microfarad = 10^{-6} F

1 picofarad = 10^{-12} F

$$C \equiv \frac{Q}{\Delta V} \tag{16.8}$$

$$1 \text{ F} \equiv 1 \text{ C/V}$$

The **capacitance of an air-filled parallel plate capacitor** is proportional to the area of the plates and inversely proportional to the separation of the plates.

$$C = \epsilon_0 \frac{A}{d} \tag{16.9}$$

$$C = \kappa \epsilon_0 \left(\frac{A}{d} \right) \tag{16.19}$$

When a **dielectric material** (or insulator) is inserted between the plates of a capacitor, the capacitance of the device increases by a factor κ. Kappa, called the dielectric constant, is dimensionless and is characteristic of a specific material.

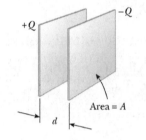

Parallel plate capacitor

The **permittivity** of free space is expressed in SI units.

$$\epsilon_0 = \frac{1}{4\pi k_e} = 8.85 \times 10^{-12} \text{ C}^2 / \text{N} \cdot \text{m}^2$$

For a **parallel combination of capacitors:**

- Each capacitor has two circuit points in common with each of the other capacitors in the group.

Capacitors in parallel

- Equivalent capacitance equals the sum of the individual capacitances.

$$C_{eq} = C_1 + C_2 + C_3 + \cdots \tag{16.12}$$

- The total charge is the sum of the charges on the individual capacitors.

$$Q_{total} = Q_1 + Q_2 + Q_3 + \cdots$$

- The potential difference is the same for each capacitor.

$$\Delta V_{total} = \Delta V_1 = \Delta V_2 = \Delta V_3 = \cdots$$

For **series combination of capacitors**:

- Adjacent capacitors have one circuit point in common.

- The reciprocal of the equivalent capacitance is equal to the sum of the reciprocals of the individual capacitances.

- The potential difference across the group equals the sum of the potential differences across each capacitor.

- The total charge on the series combination is equal to the charge on each capacitor, each capacitor having the same charge.

The **electrostatic potential energy** stored in the electric field of a charged capacitor is equal to the work done by a battery (or other source of emf) in charging the capacitor from $q = 0$ to $q = Q$.

Capacitors in series

$$\frac{1}{C_{eq}} = \frac{1}{C_1} + \frac{1}{C_2} + \frac{1}{C_3} + \dots \quad (16.15)$$

$$\Delta V = \Delta V_1 + \Delta V_2 + \Delta V_3 + \dots$$

$$Q_{total} = Q_1 = Q_2 = Q_3 = \dots$$

$$W = \tfrac{1}{2} Q \Delta V \quad (16.16)$$

$$\text{Energy stored} = \tfrac{1}{2} C(\Delta V)^2 = \frac{Q^2}{2C} \quad (16.17)$$

SUGGESTIONS, SKILLS, AND STRATEGIES

A STRATEGY FOR PROBLEMS INVOLVING ELECTRIC POTENTIAL

1. When working problems involving electric potential, remember that electric potential is a scalar quantity (rather than a vector quantity like the electric field), so there are no components to worry about. Therefore, when using the superposition principle to evaluate the electric potential at a point due to a system of point charges, you simply take the algebraic sum of the electric potentials due to each charge. However, you must keep track of signs. The electric potential due to each positive charge is positive; the electric potential due to each negative charge is negative. The basic equation to use is $V = k_e q / r$.

2. As in mechanics, only changes in potential energy are significant; hence, the point where you choose the potential energy to be zero is arbitrary. However, it is common practice to set $V = 0$ at infinity when this choice suits the particular situation.

SOLVING STRATEGY FOR CAPACITORS

1. Be careful with your choice of units. To calculate the capacitance of a device in farads, make sure that the plate separation is in meters.

2. **When two or more capacitors are connected in series**, they carry the same charge, but the potential differences across them are not the same unless they have equal

values of capacitance. The reciprocal of the equivalent capacitance is the sum of the reciprocals of the individual capacitances. The equivalent capacitance of the combination is always less than that of the smallest individual capacitor.

3. **When two or more capacitors are connected in parallel,** the potential differences across each are the same. The charge on each capacitor is proportional to its capacitance; hence, the capacitances add directly to give the equivalent capacitance of the parallel combination.

4. A complicated circuit consisting of capacitors can often be reduced to a simple circuit containing only one equivalent capacitor. To do so, examine the initial circuit and replace any capacitors in series or any in parallel using the rules of Steps 2 and 3 above. Draw a sketch of the new circuit after these changes have been made. Examine this new circuit and replace any series or parallel combinations. Continue this process until a single, equivalent capacitor is found.

5. If the charge on, or the electric potential difference across, one of the capacitors in a complicated circuit is to be found, start with the final circuit found in Step 4 and gradually work your way back through the circuits using $C = Q/\Delta V$ and the rules given in Steps 2 and 3 above. Make use of the successive sketches prepared in step 4 as the original circuit was simplified.

REVIEW CHECKLIST

*   Understand that each point in the vicinity of a charge distribution can be characterized by a scalar quantity called the electric potential, *V*; and define the quantity, electrical potential difference. (Section 16.1)

*   Calculate the electric potential difference between any two points in a uniform electric field and calculate the electric potential difference between any two points in the vicinity of a group of point charges. (Sections 16.1–16.2)

*   Calculate the electric potential energy associated with a group of point charges and define the unit of energy, the electron volt. (Section 16.3)

*   Justify the claims that (i) all points on the surface and within a charged conductor are at the same potential and (ii) the electric field within a charged conductor is zero. (Sections 16.3–16.4)

*   Define the quantity, capacitance; evaluate the capacitance of a parallel plate capacitor of given area and plate separation. (Sections 16.6–16.7)

*   Determine the equivalent capacitance of a network of capacitors in series-parallel combination and calculate the final charge on each capacitor and the potential difference across each when a known potential is applied across the combination. (Section 16.8)

*   Calculate the energy stored in a charged capacitor. (Section 16.9)

SOLUTIONS TO SELECTED END-OF-CHAPTER PROBLEMS

3. A potential difference of 90 mV exists between the inner and outer surfaces of a cell membrane. The inner surface is negative relative to the outer surface. How much work is required to eject a positive sodium ion (Na^+) from the interior of the cell?

Solution

Since the inner surface of the membrane is negative relative to the outer surface, the electric field within the membrane is directed from the outer surface toward the inner surface (recall that field lines originate on positive charges and terminate on negative charges). Thus, when the positive ion moves from the inner surface to the outer surface as it leaves the cell, it is being moved against the direction of the force exerted on it by the field. This means that its electrical potential energy is increasing, so the change in potential it undergoes is

$$\Delta V = +90 \text{ mV}$$

The work an external agent must do on the positive ion to overcome the influence of the electric field and move the ion out of the cell is given by

$$W = q(\Delta V) = (+e)(\Delta V) = \left(1.60 \times 10^{-19} \text{ C}\right)\left(+90 \times 10^{-3} \text{ V}\right)$$

or $\qquad W = 1.4 \times 10^{-20} \text{ J}$ ◊

---

10. On planet Tehar, the free-fall acceleration is the same as that on Earth, but there is also a strong downward electric field that is uniform close to the planet's surface. A 2.00-kg ball having a charge of 5.00 μC is thrown upward at a speed of 20.1 m/s. It hits the ground after an interval of 4.10 s. What is the potential difference between the starting point and the top point of the trajectory?

Solution

Since both the gravitational and electric fields this charged ball moves in may be considered uniform over the range of motion, the ball will have a constant vertical acceleration. We use the kinematics equation $y = v_{0y}t + a_y t^2 / 2$, with $y = 0$ at the time $t = t_f$, to obtain

$$a_y = -\frac{2v_{0y}}{t_f} \tag{1}$$

We also make use of Newton's second law to find

$$a_y = \frac{\Sigma F_y}{m} = \frac{-mg + qE_y}{m} \quad \text{or} \quad a_y = -\left(g - \frac{qE_y}{m}\right) \tag{2}$$

Equating the results in equations (1) and (2) gives $g - \dfrac{qE_y}{m} = \dfrac{2v_{0y}}{t_f}$ or the electric field strength is

$$E_y = \frac{m}{q}\left(g - \frac{2v_{0y}}{t_f}\right) = \frac{2.00\ \text{kg}}{5.00 \times 10^{-6}\ \text{C}}\left[9.80\ \text{m/s}^2 - \frac{2(20.1\ \text{m/s})}{4.10\ \text{s}}\right] = -1.95 \times 10^3\ \text{N/C}$$

The negative sign in this result simply means the electric field is directed downward.

Next, we use the kinematics equation $v_y^2 = v_{0y}^2 + 2a_y(\Delta y)$, with $v_y = 0$ at $(\Delta y) = (\Delta y)_{max}$, and make use of equation (1) again to find

$$(\Delta y)_{max} = \frac{0 - v_{0y}^2}{2a_y} = \frac{-v_{0y}^2}{2}\left(-\frac{t_f}{2v_{0y}}\right) = \frac{v_{0y}t_f}{4} = \frac{(20.1\ \text{m/s})(4.10\ \text{s})}{4} = 20.6\ \text{m}$$

The electrical potential difference between the starting point and highest point in the trajectory is then

$$\Delta V = V_{top} - V_{ground} = \frac{\Delta PE_e}{q} = \frac{-(\text{work done by field})}{q} = \frac{-\left[qE_y(\Delta y)_{max}\right]}{q} = -E_y(\Delta y)_{max}$$

or $\Delta V = -(-1.95 \times 10^3\ \text{N/C})(+20.6\ \text{m}) = 4.02 \times 10^4\ \text{J/C} = 40.2 \times 10^3\ \text{V} = 40.2\ \text{kV}$ ◊

---

13. (a) Find the electric potential, taking zero at infinity, at the upper right corner (the corner without a charge) of the rectangle in Figure P16.13. (b) Repeat if the 2.00-μC charge is replaced with a charge of $-2.00\ \mu$C.

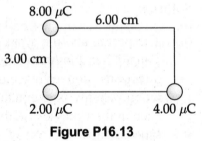

Figure P16.13

Solution

When the zero point of electric potential is at infinity, the potential at an observation point due to the presence of a collection of nearby point charges is

$$V = V_1 + V_2 + V_3 + \ldots = \Sigma V_i = \Sigma k_e \frac{q_i}{r_i}$$

where r_i is the distance from the position of the charge q_i to the observation point.

(a) For the charge configuration shown in Figure P16.13, the electric potential at the upper right corner of the rectangle is

$$V = k_e\left(\frac{q_1}{r_1} + \frac{q_2}{r_2} + \frac{q_3}{r_3}\right)$$

where $q_1 = 8.00 \ \mu C$, $r_1 = 6.00$ cm, $\quad q_2 = 4.00 \ \mu C$, $r_2 = 3.00$ cm,

$q_3 = 2.00 \ \mu C$, and $\quad r_3 = \sqrt{r_1^2 + r_2^2} = 6.71$ cm

$$V = \left(8.99 \times 10^9 \ \frac{\text{N} \cdot \text{m}^2}{\text{C}^2} \right) \left(\frac{8.00 \ \mu C}{6.00 \ \text{cm}} + \frac{4.00 \ \mu C}{3.00 \ \text{cm}} + \frac{2.00 \ \mu C}{6.71 \ \text{cm}} \right) \left(\frac{10^{-6} \ \text{C}}{1 \ \mu C} \right) \left(\frac{10^2 \ \text{cm}}{1 \ \text{m}} \right)$$

which gives $V = 2.67 \times 10^6$ V. $\qquad\qquad\qquad\qquad\qquad\qquad\qquad\qquad\qquad \lozenge$

(b) If the charge $q_3 = 2.00 \ \mu C$ is replaced by a charge $q_3' = -2.00 \ \mu C$, the new potential at the upper right corner is

$$V = \left(8.99 \times 10^9 \ \frac{\text{N} \cdot \text{m}^2}{\text{C}^2} \right) \left(\frac{8.00 \ \mu C}{6.00 \ \text{cm}} + \frac{4.00 \ \mu C}{3.00 \ \text{cm}} - \frac{2.00 \ \mu C}{6.71 \ \text{cm}} \right) \left(\frac{10^{-6} \ \text{C}}{1 \ \mu C} \right) \left(\frac{10^2 \ \text{cm}}{1 \ \text{m}} \right)$$

or $\quad V = 2.13 \times 10^6$ V $\qquad\qquad\qquad\qquad\qquad\qquad\qquad\qquad\qquad\qquad \lozenge$

---

20. A proton and an alpha particle (charge $2e$, mass 6.64×10^{-27} kg) are initially at rest, separated by 4.00×10^{-15} m. (a) If they are both released simultaneously, explain why you can't find their velocities at infinity using only conservation of energy. (b) What other conservation law can be applied in this case? (c) Find the speeds of the proton and alpha particle, respectively, at infinity.

Solution

(a) The proton and the alpha particle are both positively charged and repel one another. Thus, when they are simultaneously released from rest, they accelerate in opposite directions, converting electrical potential energy into kinetic energy. The gain in kinetic energy is unequally divided between the two particles because of the difference in their masses, so the proton and the alpha particle will have two different final speeds. However, use of conservation of energy alone, only provides one equation relating these two unknown final speeds, and will not permit a full solution. $\qquad \lozenge$

(b) A second equation involving the two unknown final speeds may be obtained by applying the principle of conservation of linear momentum. Then, one will have two equations with two unknowns, and solving these equations simultaneously will yield values for both unknowns. $\qquad\qquad\qquad\qquad\qquad\qquad\qquad\qquad\qquad\qquad \lozenge$

(c) From conservation of energy: $\qquad KE_f + PE_f = KE_i + PE_i$ we have

$$\left(\frac{1}{2} m_p v_p^2 + \frac{1}{2} m_\alpha v_\alpha^2 \right) + 0 = 0 + k_e \frac{q_p q_\alpha}{r_i} \quad \text{and} \quad m_p v_p^2 + m_\alpha v_\alpha^2 = 2 k_e \frac{q_p q_\alpha}{r_i} \quad \textbf{(1)}$$

From conservation of linear momentum:

$$m_p v_p + m_\alpha v_\alpha = 0 \quad \text{or} \quad v_\alpha = -\left(m_p / m_\alpha \right) v_p \qquad\qquad\qquad\qquad \textbf{(2)}$$

Substituting equation (2) into (1) gives $m_p v_p^2 + m_\alpha \left(m_p / m_\alpha \right)^2 v_p^2 = 2k_e \dfrac{q_p q_\alpha}{r_i}$

and reduces to $\left(m_\alpha + m_p \right) \left(\dfrac{m_p}{m_\alpha} \right) v_p^2 = 2k_e \dfrac{q_p q_\alpha}{r_i}$ or $v_p = \sqrt{ \left(\dfrac{m_\alpha}{m_p} \right) \dfrac{2k_e q_p q_\alpha}{\left(m_\alpha + m_p \right) r_i} }$

Thus, the final speed of the proton is

$$v_p = \sqrt{ \left(\frac{6.64 \times 10^{-27} \text{ kg}}{1.67 \times 10^{-27} \text{ kg}} \right) \left[\frac{2\left(8.99 \times 10^9 \text{ N} \cdot \text{m}^2 / \text{C}^2 \right)\left(1.60 \times 10^{-19} \text{ C} \right)\left(3.20 \times 10^{-19} \text{ C} \right)}{\left(6.64 \times 10^{-27} \text{ kg} + 1.67 \times 10^{-27} \text{ kg} \right)\left(4.00 \times 10^{-15} \text{ m} \right)} \right] }$$

yielding $v_p = 1.05 \times 10^7 \text{ m/s}$ ◊

Then, equation (2) gives the final speed of the alpha particle as

$$\left| v_\alpha \right| = \left(\frac{m_p}{m_\alpha} \right) v_p = \left(\frac{1.67 \times 10^{-27} \text{ kg}}{6.64 \times 10^{-27} \text{ kg}} \right)\left(1.05 \times 10^7 \text{ m/s} \right) = 2.64 \times 10^6 \text{ m/s}$$ ◊

---

29. An air-filled capacitor consists of two parallel plates, each with an area of 7.60 cm^2 and separated by a distance of 1.80 mm. If a 20.0-V potential difference is applied to these plates, calculate (a) the electric field between the plates, (b) the capacitance, and (c) the charge on each plate.

Solution

The electric field in the region between a pair of parallel plates is uniform in magnitude and direction. Hence, the force exerted by this field on a particle of charge q located in this region is constant with magnitude $F = qE$. The magnitude of the potential difference between the plates (work per unit charge required to move charge from one plate to the other) is therefore given by

$$\left| \Delta V \right| = \frac{W}{q} = \frac{Fd}{q} = \frac{(qE)d}{q} = Ed$$

where d is the distance separating the plates.

(a)   The electric field between the specified set of plates is

$$E = \frac{\left| \Delta V \right|}{d} = \frac{20.0 \text{ V}}{1.80 \times 10^{-3} \text{ m}} = 1.11 \times 10^4 \text{ V/m} = 11.1 \text{ kV/m}$$ ◊

The field is directed from the positive plate toward the negative plate.   ◊

(b) The capacitance of an air-filled parallel-plate capacitor is $C = \kappa_{air} \epsilon_0 A/d$, where $\kappa_{air} \approx 1.00$, ϵ_0 is the permittivity of free space, and A is the surface area of one of the plates. For the given set of plates,

$$C = \frac{(1.00)(8.85 \times 10^{-12} \text{ C}^2/\text{N} \cdot \text{m}^2)(7.60 \text{ cm}^2)}{1.80 \times 10^{-3} \text{ m}} \left(\frac{1 \text{ m}}{10^2 \text{ cm}} \right)^2$$

or $C = 3.74 \times 10^{-12}$ F = 3.74 pF   ◊

(c) The magnitude of the charge on each plate when the potential difference is $\Delta V = 20.0$ V is given by

$$Q = C(\Delta V) = (3.74 \times 10^{-12} \text{ F})(20.0 \text{ V}) = 7.47 \times 10^{-11} \text{ C} = 74.7 \text{ pC}$$ ◊

---

35. (a) Find the equivalent capacitance of the capacitors in Figure P16.35. (b) Find the charge on each capacitor. (c) Find the potential difference across each capacitor.

Solution

(a) The equivalent capacitance of the parallel combination between points b and c is $C_{bc} = 6.00$ μF + 2.00 μF = 8.00 μF.

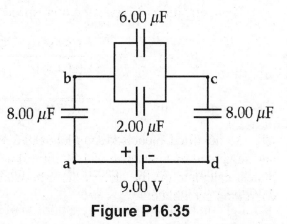

Figure P16.35

Replacing this parallel combination by its equivalent capacitance gives us three 8.00 μF capacitors connected in series between points a and d. The equivalent capacitance for the full circuit is then

$$\frac{1}{C_{eq}} = \frac{1}{8.00 \text{ } \mu\text{F}} + \frac{1}{8.00 \text{ } \mu\text{F}} + \frac{1}{8.00 \text{ } \mu\text{F}} = \frac{3}{8.00 \text{ } \mu\text{F}} \quad \text{or} \quad C_{eq} = \frac{8.00 \text{ } \mu\text{F}}{3} = 2.67 \text{ } \mu\text{F}$$ ◊

(b) Since capacitors in series all have the same charge, the charge on each capacitor in the series combination is

$$Q_{ab} = Q_{bc} = Q_{cd} = C_{eq}(\Delta V_{ad}) = (2.67 \text{ } \mu\text{F})(9.00 \text{ V}) = 24.0 \text{ } \mu\text{C}$$

Also, we find that $\Delta V_{bc} = \dfrac{Q_{bc}}{C_{bc}} = \dfrac{24.0 \text{ } \mu\text{C}}{8.00 \text{ } \mu\text{F}} = 3.00$ V

The charge on each individual capacitor is

For the 8.00 μF between points a and b: $Q_8 = Q_{ab} = 24.0 \text{ } \mu\text{C}$ ◊

For the 8.00 μF between points c and d: $Q_8 = Q_{cd} = 24.0 \text{ } \mu\text{C}$ ◊

For the 6.00 μF: $Q_6 = C_6(\Delta V_{bc}) = (6.00 \text{ } \mu\text{F})(3.00 \text{ V}) = 18.0 \text{ } \mu\text{C}$ ◊

And, for the 2.00 μF: $Q_2 = C_2(\Delta V_{bc}) = (2.00\ \mu F)(3.00\ V) = 6.00\ \mu C$ ◊

Note that $Q_2 + Q_6 = Q_{bc}$ as it must.

(c) The potential difference across each capacitor in the circuit is:

For the 8.00 μF between points a and b: $\quad \Delta V_8 = \dfrac{Q_8}{C_8} = \dfrac{24.0\ \mu C}{8.00\ \mu F} = 3.00\ V$ ◊

For the 6.00 μF and 2.00 μF in parallel: $\quad \Delta V_6 = \Delta V_2 = \Delta V_{bc} = 3.00\ V$ ◊

For the 8.00 μF between points c and d: $\quad \Delta V_8 = \dfrac{Q_8}{C_8} = \dfrac{24.0\ \mu C}{8.00\ \mu F} = 3.00\ V$ ◊

---

43. A 1.00-μF capacitor is charged by being connected across a 10.0-V battery. It is then disconnected from the battery and connected across an uncharged 2.00-μF capacitor. Determine the resulting charge on each capacitor.

Solution

The charge initially stored on the 1.00-μF capacitor by the battery is

$$Q = C_1 \Delta V_{battery} = (1.00\ \mu F)(10.0\ V) = 10.0\ \mu C$$

If this capacitor is now carefully disconnected from the battery, and then connected in parallel with an uncharged 2.00-μF, the charge Q remains stored as the total charge in the parallel combination. However, a portion of it will now be stored on the 2.00-μF capacitor with the remainder left on the 1.00-μF capacitor.

The equivalent capacitance of the parallel combination is

$$C_{eq} = C_1 + C_2 = 3.00\ \mu F$$

and, with a total stored charge of $Q = 10.0\ \mu C$, the potential difference across each capacitor in the combination is

$$\Delta V_p = \frac{Q}{C_{eq}} = \frac{10.0\ \mu C}{3.00\ \mu F} = 3.33\ V$$

The charged stored on each capacitor in the parallel combination is then

For the 1.00-μF: $\quad Q_1 = C_1 \Delta V_p = (1.00\ \mu F)(3.33\ V) = 3.33\ \mu C$ ◊

For the 2.00-μF: $\quad Q_2 = C_2 \Delta V_p = (2.00\ \mu F)(3.33\ V) = 6.67\ \mu C$ ◊

Note that $Q_1 + Q_2 = Q = 10.0\ \mu C$ as it must in order to conserve charge.

---

47. A parallel plate capacitor has capacitance 3.00 μF. (a) How much energy is stored in the capacitor if it is connected to a 6.00-V battery? (b) If the battery is disconnected and the distance between the charged plates doubled, what is the energy stored? (c) The battery is subsequently reattached to the capacitor, but the plate separation remains as in part (b). How much energy is stored? (Answer each part in micro Joules.)

Solution

(a) The energy stored in a capacitor may be expressed in two equivalent ways:

$$\text{Energy Stored} = \frac{Q^2}{2C} \quad \text{and} \quad \text{Energy Stored} = \frac{1}{2}C(\Delta V)^2$$

In this case, the potential difference across the capacitor is known and we use the latter form to compute

$$\left(\text{Energy Stored}\right)_a = \frac{1}{2}\left(3.00 \times 10^{-6} \text{ F}\right)\left(6.00 \text{ V}\right)^2 = 54.0 \times 10^{-6} \text{ J} = 54.0 \ \mu\text{J} \qquad \lozenge$$

(b) The capacitance of an air-filled parallel plate capacitor is $C = \epsilon_0 A/d$, so it is seen that doubling the distance d between the plates reduces the capacitance by a factor of 2, i.e., $C_f = C_i/2$. Once the capacitor is disconnected from the battery, the charge on it has no way to leave and must remain constant. Thus, we use the first form for the energy stored to obtain

$$\left(\text{Energy Stored}\right)_b = \frac{Q_f^2}{2C_f} = \frac{Q_i^2}{2(C_i/2)} = 2\left(\frac{Q_i^2}{2C_i}\right) = 2\left(\text{Energy Stored}\right)_a$$

or $\quad \left(\text{Energy Stored}\right)_b = 2\left(54.0 \ \mu\text{J}\right) = 108 \ \mu\text{J}$ $\qquad \lozenge$

(c) Once the battery is reconnected, with the capacitance left at value $C_f = C_i/2$, the charge will readjust itself until the potential difference across the capacitor again matches that across the battery. Then, the second expression for the energy stored gives

$$\left(\text{Energy Stored}\right)_c = \frac{1}{2}C_f(\Delta V)_f^2 = \frac{1}{2}\left(\frac{C_i}{2}\right)(\Delta V)_i^2 = \frac{1}{2}\left[\frac{1}{2}C_i(\Delta V)_i^2\right] = \frac{1}{2}\left(\text{Energy Stored}\right)_a$$

or $\quad \left(\text{Energy Stored}\right)_c = \frac{1}{2}\left(54.0 \ \mu\text{J}\right) = 27.0 \ \mu\text{J}$ $\qquad \lozenge$

51. Determine (a) the capacitance and (b) the maximum voltage that can be applied to a Teflon®-filled parallel-plate capacitor having a plate area of 175 cm^2 and an insulation thickness of $0.040\ 0$ mm.

Solution

(a) A parallel-plate capacitor with a dielectric material between the plates has a capacitance of $C = \kappa\, \epsilon_0\, A/d$, where κ is the dielectric constant of the material, ϵ_0 is the permittivity of free space, A is the area of one of the plates, and d is the distance separating the plates. From Table 16.1 in the textbook, the dielectric constant of Teflon® is $\kappa = 2.1$, so the capacitance of this parallel-plate capacitor is

$$C = \frac{(2.1)(8.85 \times 10^{-12}\ \text{C}^2/\text{N} \cdot \text{m}^2)(175 \times 10^{-4}\ \text{m}^2)}{0.040\ 0 \ \times 10^{-3}\ \text{m}} = 8.1 \times 10^{-9}\ \text{F} = 8.1\ \text{nF} \quad \Diamond$$

(b) The dielectric strength of a material is the maximum electric field that material can withstand before it breaks down and begins to conduct. From Table 16.1, the dielectric strength for Teflon® is $E_{\text{max}} = 60 \times 10^6$ V/m. The potential difference between the plates of a parallel-plate capacitor is $\Delta V = E d$, where E is the magnitude of the electric field and d is the distance between the plates. Thus, the maximum voltage that can be applied to this Teflon®-filled capacitor is

$$\Delta V_{\text{max}} = E_{\text{max}} d = (60 \times 10^6\ \text{V/m})(0.040\ 0 \times 10^{-3}\ \text{m})$$

or $\Delta V_{\text{max}} = 2.4 \times 10^3\ \text{V} = 2.4\ \text{kV}$ $\Diamond$

---

57. A parallel-plate capacitor with a plate separation d has a capacitance C_0 in the absence of a dielectric. A slab of dielectric material of dielectric constant κ and thickness $d/3$ is then inserted between the plates as in Figure P16.57. Show that the capacitance of this partially filled capacitor is given by

$$C = \left(\frac{3\kappa}{2\kappa + 1} \right) C_0$$

(Hint: Treat the system as two capacitors connected in series, one with dielectric in it and the other one empty.)

Solution

Without a dielectric present, the capacitance of the parallel plate capacitor is

$$C_0 = \epsilon_0\, A/d$$

Once the dielectric is inserted, it fills one-third of the gap between the plates as shown in part (a) of the sketch at the right. This situation can be thought of as having two capacitors,

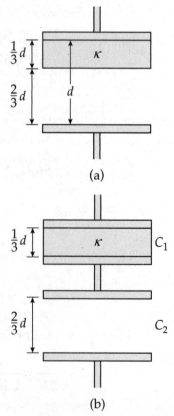

(a)

(b)

C_1 and C_2, in series as shown in part (b) of the sketch. In reality, the lower plate of C_1 and the upper plate of C_2 are the same surface, which is the lower surface of the dielectric shown in part (a) of the sketch.

The capacitances of the individual capacitors in our model are:

$$C_1 = \frac{\kappa \in_0 A}{(d/3)} = \frac{3\kappa \in_0 A}{d} \quad \text{and} \quad C_2 = \frac{\in_0 A}{(2d/3)} = \frac{3 \in_0 A}{2d}$$

The capacitance of the parallel plate capacitor after the dielectric is inserted is the equivalent capacitance of the series combination in our model of part (b) of the sketch. Thus,

$$\frac{1}{C} = \frac{1}{C_1} + \frac{1}{C_2} = \frac{d}{3\kappa \in_0 A} + \frac{2d}{3 \in_0 A} = \frac{d}{3 \in_0 A}\left(\frac{1}{\kappa} + 2\right) = \frac{d}{3 \in_0 A}\left(\frac{2\kappa + 1}{\kappa}\right)$$

which yields $C = \dfrac{3 \in_0 A}{d}\left(\dfrac{\kappa}{2\kappa + 1}\right) = \left(\dfrac{3\kappa}{2\kappa + 1}\right)\dfrac{\in_0 A}{d} = \left(\dfrac{3\kappa}{2\kappa + 1}\right)C_0$ ◊

---

62. A spherical capacitor consists of a spherical conducting shell of radius b and charge $-Q$ concentric with a smaller conducting sphere of radius a and charge Q. (a) Find the capacitance of this device. (b) Show that as the radius b of the outer sphere approaches infinity, the capacitance approaches the value $a/k_e = 4\pi \in_0 a$.

Solution

(a)   Due to spherical symmetry, the charge on each of the concentric spherical shells will be uniformly distributed over that shell. Inside a spherical surface having a uniform charge distribution, the electric field due to the charge on that surface is zero. Thus, in this region, the potential due to the charge on that surface is constant and equal to the potential at the surface. Outside a spherical surface having a uniform charge distribution, the potential due to the charge on that surface is given by $V = k_e q/r$, where r is the distance from the center of that surface, and q is the charge on that surface.

In the region **between** a pair of concentric spherical shells, with the inner shell having charge $+Q$ and the outer shell having radius b and charge $-Q$, the total electric potential is given by

$$V = V_{\substack{\text{due to} \\ \text{inner shell}}} + V_{\substack{\text{due to} \\ \text{outer shell}}} = \frac{k_e Q}{r} + \frac{k_e(-Q)}{b} = k_e Q\left(\frac{1}{r} - \frac{1}{b}\right)$$

The potential difference between the two shells is therefore,

$$\Delta V = V|_{r=a} - V|_{r=b} = k_e Q\left(\frac{1}{a} - \frac{1}{b}\right) - k_e Q\left(\frac{1}{b} - \frac{1}{b}\right) = k_e Q\left(\frac{b-a}{ab}\right)$$

The capacitance of this device is given by

$$C = \frac{Q}{\Delta V} = \frac{ab}{k_e(b-a)}$$ ◊

(b) When b is very large in comparison to a, we have $b - a \approx b$. Thus, in the limit as $b \to \infty$, the capacitance found above becomes

$$C \to \frac{ab}{k_e(b)} = \frac{a}{k_e} = 4\pi \in_0 a \qquad \Diamond$$

---

65. Capacitors $C_1 = 6.0 \ \mu F$ and $C_2 = 2.0 \ \mu F$ are charged as a parallel combination across a 250-V battery. The capacitors are disconnected from the battery and from each other. They are then connected positive plate to negative plate and negative plate to positive plate. Calculate the resulting charge on each capacitor.

Solution

When connected in parallel across the battery, the potential difference across each of the capacitors is $\Delta V_6 = \Delta V_2 = \Delta V_{battery} = 250$ V. The charge initially stored on each of the capacitors by the battery is

$$Q_6 = C_6(\Delta V_6) = (6.0 \ \mu F)(250 \text{ V}) = 1.5 \times 10^3 \ \mu C$$

and $\quad Q_2 = C_2(\Delta V_2) = (2.0 \ \mu F)(250 \text{ V}) = 5.0 \times 10^2 \ \mu C$

When the capacitors are reconnected, with the positive plate of one connected to the negative plate of the other and vice versa, the magnitude of the net charge on each side of this new parallel combination is

$$Q_{net} = Q_1 - Q_2 = 1.5 \times 10^3 \ \mu C - 5.0 \times 10^2 \ \mu C = 1.0 \times 10^3 \ \mu C$$

The equivalent capacitance of the parallel combination of these two capacitors is

$$C_{eq} = C_1 + C_1 = 6.0 \ \mu F + 2.0 \ \mu F = 8.0 \ \mu F$$

so the potential difference across the new parallel combination is

$$\Delta V' = \frac{Q_{net}}{C_{eq}} = \frac{1.0 \times 10^3 \ \mu C}{8.0 \ \mu F} = 125 \text{ V}$$

and the final charge stored on each of the capacitors is

$$Q_6' = C_6(\Delta V') = (6.0 \ \mu F)(125 \text{ V}) = 7.5 \times 10^2 \ \mu C = 0.75 \text{ mC} \qquad \Diamond$$

and $\quad Q_2' = C_2(\Delta V') = (2.0 \ \mu F)(125 \text{ V}) = 2.5 \times 10^2 \ \mu C = 0.25 \text{ mC} \qquad \Diamond$

17

Current and Resistance

NOTES FROM SELECTED CHAPTER SECTIONS

17.1 Electric Current

The direction of conventional current is designated as the direction in which positive charges would flow. *In an ordinary metal conductor, the direction of conventional current will be opposite the direction of flow of electrons (which are the charge carriers in this case).*

17.2 A Microscopic View: Current and Drift Speed

In the classical model of electronic conduction in a metal, electrons are treated like molecules in a gas. In the absence of an electric field, electrons have an average velocity of zero.

Under the influence of an electric field, the electrons move along a direction opposite the direction of the applied field with a drift velocity, v_d. The drift velocity of electrons in a conductor is proportional to the current and inversely proportional to the number of free electrons per unit volume, n. *The drift speed is much smaller than the average speed of electrons between collisions.*

17.4 Resistance, Resistivity, and Ohm's Law

When a potential difference, ΔV, is applied across the ends of a metallic conductor, the current in the conductor is found to be proportional to the applied voltage. For many materials, including most metals, the resistance is constant over a wide range of applied voltages and $\Delta V = IR$. This statement is known as Ohm's law and R is called the resistance of the conductor.

Ohm's law is an empirical relationship that is valid only for certain materials. Materials which obey Ohm's law over a wide range of voltages are said to be ohmic.

Resistance has the SI units of volts per ampere, called ohms (Ω); if a potential difference of 1 V across a conductor produces a current of 1 A, the resistance of the conductor is 1 Ω.

A current can be established in a conductor by applying a potential difference between two points in the conductor (for example, between the ends of a wire). The average velocity of electrons moving through the conductor is limited by collisions with atoms of the conducting material. The overall effect of the conductor inhibiting electron flow is due to the **resistivity** (ρ) of the conducting material, which depends on the molecular structure and temperature of the conductor. *Resistivity is characteristic of a particular **type** of material (for example, copper), and the value of ρ for a material does not depend on the size or shape of the material.* Electrical resistance, R, is associated with a particular **sample** of material (for example, a cylinder of copper of specific length and cross-sectional area). Good conductors have low values of ρ.

17.5 Temperature Variation of Resistance

The manner in which resistivity of a material (and resistance of a conductor) changes with temperature depends on a parameter called the temperature coefficient of resistance, α. For most metals, α is positive and the resistivity changes approximately linearly with temperature over a limited temperature range; α has a negative value for most semiconductors.

EQUATIONS AND CONCEPTS

Electric current is defined as the rate at which charge flows through a cross section of a conductor. *Under the influence of an electric field, electric charges will move through conducting gases, liquids, and solids.* The SI unit of current is the ampere (A).

$$I = \frac{\Delta Q}{\Delta t} \tag{17.1}$$

$$1 \text{ A} = 1 \text{ C/s}$$

The **average current** in a conductor can be related to microscopic quantities in the conductor: the number of charge carriers (positive or negative) per unit volume (n), the quantity of charge (q) associated with each carrier, and the drift velocity (v_d) of the carriers.

$$I = nqv_dA \tag{17.2}$$

n = the number of mobile charge carriers per unit volume of conductor

In **Ohm's law,** as expressed in Equation (17.3), R is understood to be independent of ΔV. This form of Ohm's law is valid only for ohmic materials which have a linear current-voltage relationship over a wide range of applied voltages.

$$R = \frac{\Delta V}{I} \tag{17.3}$$

The **resistance, R, of a given conductor** made of a homogeneous material of uniform cross section, A, can be expressed in terms of the dimensions of the conductor and an intrinsic property of the material of which the conductor is made called its **resistivity.** *The value of the resistivity of a given material depends on the electronic structure of the material and on the temperature.*

$$R = \rho\frac{\ell}{A} \tag{17.5}$$

R = resistance

ρ = resistivity

The **ohm** is the SI unit of resistance. If a potential difference of 1 V across a conductor produces a current of 1 A, the resistance of the conductor is 1 ohm. Resistivity is expressed in SI units of ohm-meters ($\Omega\cdot$m).

$$1 \ \Omega = 1 \text{ V/A}$$

The **temperature coefficient of resistivity** determines the rate at which the resistivity of a material (and therefore the resistance of a conductor) varies with temperature. Over a limited range of temperatures, this variation is approximately linear. *Semiconductors (for example, carbon) are characterized by a negative temperature coefficient of resistivity, and in these materials the resistance decreases as the temperature increases.*

$$\rho = \rho_0[1 + \alpha(T - T_0)] \tag{17.6}$$

$$R = R_0[1 + \alpha(T - T_0)] \tag{17.7}$$

α = temperature coefficient of resistivity

T_0 is a reference temperature
(usually taken to be 20.0 °C)

Power delivered to a resistor or other current-carrying device can be expressed in alternative forms as shown in Equations (17.8) and (17.9).

$$\mathcal{P} = I\Delta V \tag{17.8}$$

$$\mathcal{P} = I^2 R = \frac{(\Delta V)^2}{R} \tag{17.9}$$

The **SI unit of power** is the watt (W).

$$1\text{ W} = 1\text{ J/s} = 1\text{ V} \cdot \text{A}$$

The **kilowatt-hour** is the quantity of energy consumed in one hour at a constant use rate (or power) of 1 kW. It is the unit used by electric companies to bill for energy consumption.

$$1\text{ kW h} = 3.60 \times 10^6 \text{ J} \tag{17.10}$$

REVIEW CHECKLIST

- Define the term "electric current" in terms of rate of charge flow and its corresponding unit of measure, the ampere. Calculate electron drift velocity in a conductor of specified characteristics carrying a current, *I*. (Sections 17.1–17.2)

- Sketch a simple single-loop circuit to illustrate the use of basic circuit element symbols and direction of conventional current. (Section 17.3)

- Determine the resistance of a conductor using Ohm's law. Also, calculate the resistance based on the physical characteristics of a conductor. Distinguish between ohmic and nonohmic conductors. (Section 17.4)

- Make calculations of the variation of resistance with temperature, which involves the concept of the temperature coefficient of resistivity. (Section 17.5)

- Calculate the power dissipated in a resistor. (Section 17.6)

SOLUTIONS TO SELECTED END-OF-CHAPTER PROBLEMS

1. If a current of 80.0 mA exists in a metal wire, how many electrons flow past a given cross-section of the wire in 10.0 min? In what direction do the electrons travel with respect to the current?

Solution

The current in a conductor is defined as $\quad I = \dfrac{\Delta Q}{\Delta t}$

where ΔQ is the amount of charge passing a fixed cross-section of the conductor in time Δt. Thus, if $I = 80.0$ mA $= 80.0 \times 10^{-3}$ A $= 80.0 \times 10^{-3}$ C/s, the magnitude of the charge passing a fixed cross-section of the wire in time $\Delta t = 10.0$ min is

$$|\Delta Q| = I(\Delta t) = \left(80.0 \times 10^{-3} \text{ C/s}\right)(10.0 \text{ min})\left(\frac{60.0 \text{ s}}{1 \text{ min}}\right) = 48.0 \text{ C}$$

The number of electrons that has a total charge of this magnitude is

$$N = \frac{|\Delta Q|}{e} = \frac{48.0 \text{ C}}{1.60 \times 10^{-19} \text{ C/electron}} = 3.00 \times 10^{20} \text{ electrons} \qquad \lozenge$$

The direction of conventional current is the direction that positive charges would flow through the conductor if they were free to move. That is, the current is in the direction of the electric field within the conductor. Free electrons, being negative charges, experience an electrical force and move in the direction opposite to that of the electric field, and hence opposite to the direction of the conventional current. A sketch, showing the relative directions of the current and the electrons' motion, is given below: $\qquad \lozenge$

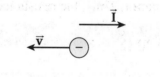

7. A 200-km-long high-voltage transmission line 2.0 cm in diameter carries a steady current of 1 000 A. If the conductor is copper with a free charge density of 8.5×10^{28} electrons per cubic meter, how many years does it take one electron to travel the full length of the cable?

Solution

The drift speed of charge carriers in a conductor carrying current I is

$$v_d = \frac{I}{nqA}$$

where n is the number of mobile charge carriers per unit volume, q is the magnitude of the charge of each carrier, and A is the cross-sectional area of the conductor.

The transmission line has a diameter of 2.0 cm, so its radius is $r = 1.0$ cm $= 1.0 \times 10^{-2}$ m, and the cross-sectional area is

$$A = \pi r^2 = \pi \left(1.0 \times 10^{-2} \text{ m}\right)^2 = \pi \times 10^{-4} \text{ m}^2$$

In metallic conductors, the mobile charge carriers are electrons with charge of magnitude $q = e = 1.60 \times 10^{-19}$ C, and the density of free electrons in this copper material is $n = 8.5 \times 10^{28}$ per cubic meter. Thus, when this cable carries a current $I = 1\,000$ A, the drift speed of the electrons is

$$v_d = \frac{1\,000 \text{ C/s}}{\left(8.5 \times 10^{28} \text{ m}^{-3}\right)\left(1.60 \times 10^{-19} \text{ C}\right)\left(\pi \times 10^{-4} \text{ m}^2\right)} = 2.34 \times 10^{-4} \text{ m/s}$$

and the time the electrons require to drift the full 200-km length of the cable is

$$\Delta t = \frac{L}{v_d} = \frac{200 \times 10^3 \text{ m}}{2.34 \times 10^{-4} \text{ m/s}} = \left(8.5 \times 10^8 \text{ s}\right)\left(\frac{1 \text{ yr}}{3.156 \times 10^7 \text{ s}}\right) = 27 \text{ yr} \qquad \lozenge$$

15. A potential difference of 12 V is found to produce a current of 0.40 A in a 3.2-m length of wire with a uniform radius of 0.40 cm. What is (a) the resistance of the wire? (b) the resistivity of the wire?

Solution

(a) According to Ohm's law, the potential difference across a conductor is directly proportional to the current flowing through that conductor. The proportionality constant is the resistance of the conductor. Thus, the resistance of this wire is

$$R = \frac{\Delta V}{I} = \frac{12 \text{ V}}{0.40 \text{ A}} = 30 \ \Omega \qquad \lozenge$$

(b) The resistance of a conductor is directly proportional to the conductor's length L and inversely proportional to its cross-sectional area A, where the proportionality constant, ρ, is called the resistivity of the material. That is, $R = \rho L/A$. The resistivity of the wire is then

$$\rho = \frac{RA}{L} = \frac{R\left(\pi r^2\right)}{L} = \frac{(30 \ \Omega)\pi\left(0.40 \times 10^{-2} \text{ m}\right)^2}{3.2 \text{ m}} = 4.7 \times 10^{-4} \ \Omega \cdot \text{m} \qquad \lozenge$$

19.   A wire of initial length L_0 and radius r_0 has a measured resistance of 1.0 Ω. The wire is drawn under tensile stress to a new uniform radius of $r = 0.25 r_0$. What is the new resistance of the wire?

Solution

The volume of material making up the wire is constant as the wire is placed under stress and drawn to have a new length with a new cross-sectional area. This constant volume is

$$V = \left(\pi r^2\right)L = \left(\pi r_0^2\right)L_0$$

Thus, if the final radius of the wire is $r = 0.25 r_0 = r_0/4$, the final length of the wire must be

$$L = L_0\left(\frac{r_0^2}{r^2}\right) = L_0\left(\frac{r_0}{r}\right)^2 = L_0\left(\frac{r_0}{0.25\, r_0}\right)^2 = L_0(4)^2 = 16 L_0$$

The resistance of a cylindrical conductor is given by

$$R = \rho\frac{L}{A} = \rho\frac{L}{\left(\pi r^2\right)}$$

so the new resistance of the wire is

$$R = \rho\frac{L}{\left(\pi r^2\right)} = \rho\frac{16 L_0}{\left[\pi\left(r_0/4\right)^2\right]} = \rho\frac{256 L_0}{\left(\pi r_0^2\right)} = 256\left[\rho\frac{L_0}{\left(\pi r_0^2\right)}\right] = 256 R_0$$

or $\qquad R = 256(1.0\ \Omega) = 256\ \Omega$ ◊

---

23.   While taking photographs in Death Valley on a day when the temperature is 58.0°C, Bill Hiker finds that a certain voltage applied to a copper wire produces a current of 1.000 A. Bill then travels to Antarctica and applies the same voltage to the same wire. What current does he register there if the temperature is –88.0°C? Assume that no change occurs in the wire's shape and size.

Solution

The same applied voltage, ΔV, is used in the initial test (in Death Valley) and in the final test (conducted in Antarctica). Therefore, from Ohm's law,

$$\Delta V = I_{DV}R_{DV} = I_A R_A \quad \text{and} \quad I_A = \left(\frac{R_{DV}}{R_A}\right)I_{DV} \tag{1}$$

At temperature T, the resistance of a conducting element is given by

$$R = R_0\left[1 + \alpha\left(T - T_0\right)\right]$$

where α is the temperature coefficient of resistivity for the material in use, and R_0 is the resistance of that material at the reference temperature (normally 20.0°C).

For copper, Table 17.1 of the textbook gives $\alpha = 3.90 \times 10^{-3} (°C)^{-1}$, so Equation (1) gives the current in Antarctica as

$$I_A = \left(\frac{1 + \alpha(T_{DV} - T_0)}{1 + \alpha(T_A - T_0)} \right) I_{DV}$$

or $$I_A = \left(\frac{1 + \left[3.90 \times 10^{-3} (°C)^{-1} \right](58.0°C - 20.0°C)}{1 + \left[3.90 \times 10^{-3} (°C)^{-1} \right](-88.0°C - 20.0°C)} \right) (1.000 \text{ A}) = 1.98 \text{ A}$$ ◊

32. A platinum resistance thermometer has resistances of 200.0 Ω when placed in a 0°C ice bath and 253.8 Ω when immersed in a crucible containing melting potassium. What is the melting point of potassium? [*Hint:* First determine the resistance of the platinum resistance thermometer at room temperature, 20°C.]

Solution

The resistance of a conductor varies with temperature according to the relation $R = R_0 \left[1 + \alpha(T - T_0) \right]$, where R_0 is the resistance at the reference temperature T_0 (normally 20.0°C), and α is a property of the conducting material called the temperature coefficient of resistivity. For platinum, Table 17.1 in the textbook gives $\alpha = 3.92 \times 10^{-3} (°C)^{-1}$.

Thus, if the thermometer has resistance $R = 200.0$ Ω at 0°C, its resistance at the reference temperature of $T_0 = 20.0$°C is

$$R_0 = \frac{R}{1 + \alpha(T - T_0)} = \frac{200.0 \text{ Ω}}{1 + \left[3.92 \times 10^{-3} (°C)^{-1} \right](0°C - 20.0°C)} = 217 \text{ Ω}$$

and the temperature at which it has a resistance of $R = 253.8$ Ω (and hence, the temperature of the melting potassium) is

$$T = T_0 + \frac{(R/R_0) - 1}{\alpha} = 20.0°C + \frac{(253.8 \text{ Ω}/217 \text{ Ω}) - 1}{3.92 \times 10^{-3} (°C)^{-1}} = 63.3°C$$ ◊

37. The heating element of a coffeemaker operates at 120 V and carries a current of 2.00 A. Assuming that the water absorbs all of the energy converted by the resistor, calculate how long it takes to heat 0.500 kg of water from room temperature (23.0°C) to the boiling point.

Solution

The energy required to raise the temperature of the water from $T_i = 23.0°C$ to the boiling point $(T = 100°C)$ is

$$Q = mc(\Delta T) = (0.500 \text{ kg})(4\,186 \text{ J/kg} \cdot °C)(100°C - 23.0°C) = 1.61 \times 10^5 \text{ J}$$

The rate at which the heating element is converting electrical potential energy into internal energy of the water is

$$\mathcal{P} = (\Delta V)I = (120 \text{ V})(2.00 \text{ A}) = 240 \text{ W} = 240 \text{ J/s}$$

Thus, the time required to bring the water to a boil will be

$$\Delta t = \frac{Q}{\mathcal{P}} = \frac{1.61 \times 10^5 \text{ J}}{240 \text{ J/s}} = (671 \text{ s})\left(\frac{1 \text{ min}}{60.0 \text{ s}}\right) = 11.2 \text{ min} \qquad \Diamond$$

---

41. A copper cable is designed to carry a current of 300 A with a power loss of 2.00 W/m. What is the required radius of this cable?

Solution

The rate at which electrical power is dissipated when current flows through a resistance is given by

$$\mathcal{P} = I^2 R$$

Thus, if the cable is to have a power loss per unit length of 2.00 W/m when carrying a current of $I = 300$ A, the resistance per unit length of the cable must be

$$\frac{R}{L} = \frac{\mathcal{P}/L}{I^2} = \frac{2.00 \text{ W/m}}{(300 \text{ A})^2} = 2.22 \times 10^{-5} \text{ } \Omega/\text{m}$$

From $R = \rho \dfrac{L}{A} = \rho \dfrac{L}{\pi r^2}$ and Table 17.1 of the textbook, the required radius of the copper cable is

$$r = \sqrt{\frac{\rho}{\pi(R/L)}} = \sqrt{\frac{1.7 \times 10^{-8} \text{ } \Omega \cdot \text{m}}{\pi(2.22 \times 10^{-5} \text{ } \Omega/\text{m})}} = 1.6 \times 10^{-2} \text{ m} = 1.6 \text{ cm} \qquad \Diamond$$

---

46. An office worker uses an immersion heater to warm 250 g of water in a light, covered, insulated cup from 20°C to 100°C in 4.00 minutes. The heater is a Nichrome resistance wire connected to a 120-V power supply. Assume that the wire is at 100°C throughout the 4.00-min time interval. Specify a diameter and a length that the wire can have. Can it be made from less than 0.5 cm³ of Nichrome?

Solution

The power dissipation required to warm the water in the specified time is

$$P = \frac{\Delta Q}{\Delta t} = \frac{mc(\Delta T)}{\Delta t} = \frac{(0.250 \text{ kg})(4\,186 \text{ J/kg} \cdot °\text{C})(100°\text{C} - 20°\text{C})}{240 \text{ s}} = 3.5 \times 10^2 \text{ W}$$

Since the power dissipation by a resistor may be expressed as $P = (\Delta V)^2/R$, the needed resistance at the operating temperature of 100°C is

$$R = \frac{(\Delta V)^2}{P} = \frac{(120 \text{ V})^2}{3.5 \times 10^2 \text{ W}} = 41 \text{ } \Omega$$

Using $R = R_0\left[1 + \alpha(T - T_0)\right]$, with $\alpha = 0.40 \times 10^{-3} (°\text{C})^{-1}$, from Table 17.1 in the textbook, the resistance at $T_0 = 20°$C is found to be

$$R_0 = \frac{R}{1 + \alpha(T - T_0)} = \frac{41 \text{ } \Omega}{1 + \left[0.40 \times 10^{-3}(°\text{C})^{-1}\right](100°\text{C} - 20°\text{C})} = 40 \text{ } \Omega$$

Then, with a diameter d, cross-sectional area $A = \pi d^2/4$, and length L, the resistance at 20°C is given by $R_0 = \rho_0 L/A = 4\rho_0 L/\pi d^2$, where $\rho_0 = 150 \times 10^{-8} \text{ } \Omega \cdot \text{m}$ (for Nichrome from Table 17.1). Thus, the relation between the length and diameter of the wire must be

$$L = \left(\frac{\pi R_0}{4\rho_0}\right) d^2 = \left[\frac{\pi(40 \text{ } \Omega)}{4(150 \times 10^{-8} \text{ } \Omega \cdot \text{m})}\right] d^2 = (2.1 \times 10^7 \text{ m}^{-1}) d^2$$

Any combination of length and diameter satisfying the relation $L = (2.1 \times 10^7 \text{ m}^{-1}) d^2$ will be satisfactory. A typical combination might be $d = 0.40 \text{ mm} = 4.0 \times 10^{-4} \text{ m}$ and

$$L = (2.1 \times 10^7 \text{ m}^{-1})(4.0 \times 10^{-4} \text{ m})^2 = 3.4 \text{ m} \qquad \Diamond$$

Such a heating element could easily be made from less than 0.5 cm³ of Nichrome. The typical wire described above has a volume of

$$V = \left(\frac{\pi d^2}{4}\right) L = \frac{\pi}{4}(4.0 \times 10^{-4} \text{ m})^2 (3.4 \text{ m}) = 4.3 \times 10^{-7} \text{ m}^3 \left(\frac{10^6 \text{ cm}^3}{1 \text{ m}^3}\right) = 0.43 \text{ cm}^3 \qquad \Diamond$$

53. A particular wire has a resistivity of 3.0×10^{-8} $\Omega \cdot$ m and a cross-sectional area of 4.0×10^{-6} m². A length of this wire is to be used as a resistor that will develop 48 W of power when connected across a 20-V battery. What length of wire is required?

Solution

As the moving charges constituting a current I undergo a decrease in electric potential of ΔV, the power dissipated is $P = (\Delta V)I$. For an ohmic resistance, we may use Ohm's law and write this as

$$P = (\Delta V)I = (IR)I = I^2 R \quad \text{or} \quad P = (\Delta V)I = (\Delta V)(\Delta V/R) = (\Delta V)^2/R$$

The resistance of length L of a wire having resistivity ρ and cross-sectional area A is $R = \rho L/A$. Therefore, the last expression for the power dissipated becomes

$$P = \frac{(\Delta V)^2}{R} = \frac{A(\Delta V)^2}{\rho L}$$

If the specified wire is to dissipate 48 W of power when connected to a 20-V battery, its length must be

$$L = \frac{A(\Delta V)^2}{\rho P} = \frac{(4.0 \times 10^{-6} \text{ m}^2)(20 \text{ V})^2}{(3.0 \times 10^{-8} \text{ W} \cdot \text{m})(48 \text{ W})} = 1.1 \times 10^3 \text{ m} = 1.1 \text{ km} \qquad \lozenge$$

---

57. You are cooking breakfast for yourself and a friend using a 1 200-W waffle iron and a 500-W coffeepot. Usually, you operate these appliances from a 110-V outlet for 0.500 h each day. (a) At 12 cents per kWh, how much do you spend to cook breakfast during a 30.0 day period? (b) You find yourself addicted to waffles and would like to upgrade to a 2 400-W waffle iron that will enable you to cook twice as many waffles during a half-hour period, but you know that the circuit breaker in your kitchen is a 20-A breaker. Can you do the upgrade?

Solution

(a)   The total power used by the two appliances is

$$P_{\text{total}} = P_{\substack{\text{waffle} \\ \text{iron}}} + P_{\substack{\text{coffee} \\ \text{maker}}} = 1\,200 \text{ W} + 500 \text{ W} = 1\,700 \text{ W} = 1.70 \text{ kW}$$

The total energy used making breakfast during a 30.0 day period is then

$$E = P \cdot t = (1.70 \text{ kW})\left[(0.500 \text{ h/d})(30.0 \text{ d})\right] = 25.5 \text{ kWh}$$

and the cost is

$$cost = E \cdot rate = (25.5 \text{ kWh})(\$0.12/\text{kWh}) = \$3.06 \qquad \lozenge$$

(b) With the upgraded waffle maker, the total power requirement would be

$$\mathcal{P}_{total} = \mathcal{P}_{\substack{waffle \\ iron}} + \mathcal{P}_{\substack{coffee \\ maker}} = 2\,400 \text{ W} + 500 \text{ W} = 2\,900 \text{ W} = 2.90 \text{ kW}$$

At $\Delta V = 110$ V, the current needed to supply this much power is

$$I = \frac{\mathcal{P}_{total}}{\Delta V} = \frac{2\,900 \text{ W}}{110 \text{ V}} = 26.4 \text{ A}$$

which exceeds the 20-A limit of your circuit breaker. Thus, the present electrical service in your kitchen will not permit the upgrade. ◊

---

65. An x-ray tube used for cancer therapy operates at 4.0 MV, with a beam current of 25 mA striking the metal target. Nearly all the power in the beam is transferred to a stream of water flowing through holes drilled in the target. What rate of flow, in kilograms per second, is needed if the rise in temperature (ΔT) of the water is not to exceed 50°C?

Solution

The power used by the x-ray tube and transferred to the target by the beam is

$$\mathcal{P} = (\Delta V)I = (4.0 \times 10^6 \text{ V})(25 \times 10^{-3} \text{ A}) = 1.0 \times 10^5 \text{ W} = 1.0 \times 10^5 \text{ J/s}$$

The energy deposited in the target each second is

$$\Delta Q = \mathcal{P} \cdot t = (1.0 \times 10^5 \text{ J/s})(1.0 \text{ s}) = 1.0 \times 10^5 \text{ J}$$

When this energy is transferred to a mass m of water, the water will experience a temperature increase described by the relation $\Delta Q = mc_{water}(\Delta T)$. If the increase in temperature is to be $\Delta T < 50$°C, the minimum mass of water needed to flow through the cooling channels in the target each second is

$$m_{min} = \frac{\Delta Q}{c_{water}(\Delta T)_{max}} = \frac{1.0 \times 10^5 \text{ J}}{(4\,186 \text{ J/kg} \cdot \text{°C})(50\text{°C})} = 0.48 \text{ kg} \qquad ◊$$

18

Direct-Current Circuits

NOTES FROM SELECTED CHAPTER SECTIONS

18.1   Sources of EMF

A source of emf (for example, a battery or generator) maintains the current in a closed circuit. The emf of the source is the work done per unit charge in increasing the electric potential energy of the circulating charges. *One joule of work is required to move one coulomb of charge between two points which differ in potential by one volt.*

The terminal voltage (ΔV between the positive and negative terminals of a battery or other source) is equal to the emf of the battery when the current in the circuit is zero. In this case the terminal voltage is called the **open circuit voltage.** When the battery is delivering current to a circuit, the terminal voltage **(closed circuit voltage)** is less than the emf. This is due to the internal resistance of the battery.

18.2   Resistors in Series

For a group of resistors connected in series:

- The equivalent resistance of the series combination is the algebraic sum of the individual resistances. See Equation (18.4).

- The current is the same for each resistor in the group.

- The total potential difference across the group of resistors equals the sum of the potential differences across the individual resistors.

18.3   Resistors in Parallel

For a group of resistors connected in parallel:

- The reciprocal of the equivalent resistance of the group is the algebraic sum of the reciprocals of the individual resistors. See Equation (18.6).

- The potential differences across the individual resistors have the same value.

- The total current associated with the parallel group is the sum of the currents in the individual resistors.

18.4   Kirchhoff's Rules and Complex DC Circuits

When all resistors in a circuit are connected in series and (or) parallel combinations, Ohm's law and the properties stated above in Sections 18.2 and 18.3 are adequate to analyze the circuit to determine the current in each resistor and the potential difference between any two points in the circuit.

When circuits are formed so that they cannot be reduced to a single equivalent resistor, it is necessary to use Kirchhoff's rules to analyze the circuit.

1. **Junction rule** (**conservation of charge**): The sum of the currents entering any junction must equal the sum of the currents leaving that junction. (A junction is any point in the circuit where the current can split.)

2. **Loop rule** (**conservation of energy**): The algebraic sum of the changes in potential around any closed circuit loop must be zero.

18.5 *RC* Circuits

When the switch in a dc circuit consisting only of a battery and a resistor is moved from the "open" to the "closed" position, the current in the circuit will have a constant value. Consider the case of a circuit which includes a battery, resistor, and capacitor. When the switch in this circuit is moved from the "open" to the "closed" position (the capacitor will begin charging), the charge on the capacitor increases with time, the current decreases exponentially, and the charge on the capacitor approaches its maximum value of $Q = C\mathcal{E}$ as the time, t, is allowed to go to infinity.

When the switch in a circuit containing an initially charged capacitor and a resistor is moved from the "open" to the "closed" position (the capacitor will begin discharging) the charge on the capacitor, current in the circuit, and the voltage across the resistor will all decrease exponentially with time.

In the process of charging or discharging a capacitor, the potential difference between the plates changes as charges are transferred from one plate of the capacitor to the other. *The transfer of charge produces a current in the circuit; the charges do not move across the gap between the plates of the capacitor.*

EQUATIONS AND CONCEPTS

The **terminal voltage of a battery** will be less than the emf when the battery is providing a current to an external circuit. This is due to **internal resistance** of the battery. The terminal voltage equals the emf when the current is zero. *Often the internal resistance, r, of the source is very small compared to the external load resistance, R, of the circuit, and r can be neglected.*

$$\Delta V = \mathcal{E} - Ir \qquad (18.1)$$

$$I = \frac{\mathcal{E}}{R + r}$$

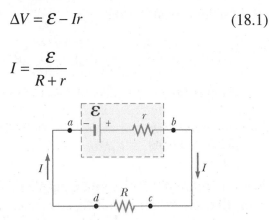

Battery with internal resistance

For resistors connected in series:

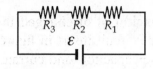

- There is only one common circuit point per pair.

- The total or equivalent resistance of a series combination of resistors is equal to the sum of the resistances of the individual resistors.

$$R_{eq} = R_1 + R_2 + R_3 + \ldots \qquad (18.4)$$

- Each resistor has the same current.

$$I_1 = I_2 = I_3 = \ldots$$

- The total potential difference equals the sum of the individual potential differences.

$$\Delta V = \Delta V_1 + \Delta V_2 + \Delta V_3 + \ldots$$

For resistors connected in parallel:

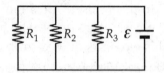

- Each resistor has two circuit points in common with each of the other resistors.

- The reciprocal of the equivalent resistance is the algebraic sum of the reciprocals of the individual resistances.

$$\frac{1}{R_{eq}} = \frac{1}{R_1} + \frac{1}{R_2} + \frac{1}{R_3} \ldots \qquad (18.6)$$

- There is a common potential difference across the group of resistors.

$$\Delta V = \Delta V_1 = \Delta V_2 = \Delta V_3 = \ldots$$

- The total current equals the sum of the currents in the individual resistors.

$$I_{total} = I_1 + I_2 + I_3 + \ldots$$

When a **capacitor is charging in series with a resistor** (see circuit in figure), the charge on the capacitor increases as given by Equation (18.7) and shown in the figure.

$$q = Q\left(1 - e^{-t/RC}\right) \qquad (18.7)$$

$$\tau = RC \qquad (18.8)$$

The **time constant** of the circuit, τ, represents the time required for the charge on the capacitor to reach 63.2% of maximum value. *Also, during one time constant, the charging current decreases from its initial maximum value of $I_i = \mathcal{E}/R$ to 36.8% of I_i.*

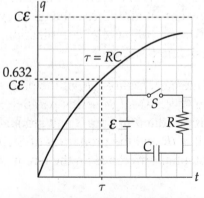

Capacitor charging in an RC circuit

When a **capacitor is discharged** through a resistor (see circuit insert in figure), the charge on the capacitor (and current) will decrease exponentially with time from an initial value of Q_0/RC to 0.368 of the initial value in a time $t = \tau$.

$$q = Qe^{-t/RC} \qquad (18.9)$$

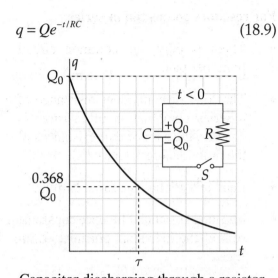

Capacitor discharging through a resistor

SUGGESTIONS, SKILLS, AND STRATEGIES

PROBLEM-SOLVING STRATEGY FOR RESISTORS

1. When two or more unequal resistors are connected in series, they carry the same current, but the potential differences across them are not the same. The resistances add directly to give the equivalent resistance of the series combination.

2. When two or more resistors (whether or not of equal value) are connected in parallel, the potential differences across them are the same. Since the current is inversely proportional to the resistance, the currents through them are not the same (except when the resistors are of equal value). *The equivalent resistance of a parallel combination of resistors is found through reciprocal addition, and the equivalent resistance is always less than the smallest individual resistance.*

3. A circuit consisting of several resistors, in series-parallel combination, can often be reduced to a simple circuit containing only one equivalent resistor. To do so, examine the initial circuit and replace any resistors in series or any in parallel using Equations (18.4) and (18.6) along with the properties outlined in Steps 1 and 2. Sketch the new circuit after these changes have been made. Examine the new circuit and replace any series or parallel combinations. **Continue this process until a single equivalent resistance is found.**

4. If the current through, or the potential difference across, a resistor in a circuit consisting of resistors in series-parallel combination is to be identified, start with the final circuit found in Step 3 and gradually work your way back through the circuits, using $\Delta V = IR$ and the properties of Steps 1 and 2. Record calculated values of current and potential difference on the circuit sketches as you proceed.

STRATEGY FOR USING KIRCHHOFF'S RULES

1. First, draw the circuit diagram and assign labels and symbols to all the known and unknown quantities. **You must assign directions to the currents in each branch of the circuit.** Do not be alarmed if you guess the direction of a current incorrectly; the resulting value will be negative, but its magnitude will be correct. Although the assignment of current directions is arbitrary, you must stick with your choice throughout as you apply Kirchhoff's rules.

2. Apply the junction rule to any junction in the circuit. The junction rule may be applied as many times as a new current (one not used in a previous application) appears in the resulting equation. **In general, the number of times the junction rule can be used is one fewer than the number of junction points in the circuit.**

3. Now apply Kirchhoff's loop rule to as many loops in the circuit as are needed to solve for the unknowns. Remember you must have as many equations as there are unknowns (I's, R's, and $\mathcal{E}$'s). **For each loop, you must first choose a direction to sum the voltage changes, clockwise or counterclockwise.** Then, you must correctly identify the increase or decrease in potential as you cross each resistor or source of emf in the closed loop.

4. Finally, you must solve the equations simultaneously for the unknown quantities. Be careful in your algebraic steps, and check your numerical answers for consistency.

Rules which should be used in determining the increase or decrease in potential difference as resistors or sources of emf are crossed in a circuit loop are stated below. *In the accompanying figures, assume that the direction of travel on the section of circuit shown is from point "a" to point "b" (left to right).*

1. The potential decreases (changes by $-IR$) when a resistor is traversed in the direction of the current.

$$V_b - V_a = -IR$$

2. The potential increases (changes by $+IR$) when a resistor is traversed in the direction opposite the direction of the current.

$$V_b - V_a = +IR$$

3. The potential increases by $+\mathcal{E}$ when a source of emf is traversed in the direction of the emf (from $-$ to $+$).

$$V_b - V_a = \mathcal{E}$$

4. The potential decreases by $-\mathcal{E}$ when a source of emf is traversed opposite to the direction of the emf (from $+$ to $-$).

$$V_b - V_a = -\mathcal{E}$$

An illustration of the use of Kirchhoff's rules for a three-loop circuit. In this illustration, the actual circuit elements, R's and $\mathcal{E}$'s, are not shown but assumed to be known. There are six possible different values of I in the circuit; therefore, you will need six independent equations to solve for the six values of I. There are four junction points in the circuit (at points a, d, f, and h). The first rule applied at **any three of these points** will yield three equations. The circuit can be thought of as a group of three "blocks" as shown in the figure. Kirchhoff's second law, when applied to each of these loops ($abcda$, $ahfga$, and $defhd$), will yield three additional equations. You can then solve the total of six equations simultaneously for the six values of I_1, I_2, I_3, I_4, I_5, and I_6 (or some combination of currents, resistances, and emf's totaling six quantities).

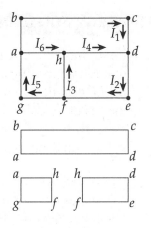

The sum of the changes in potential difference around any other closed loop in the circuit will be zero (for example, $abcdefga$ or $ahfedcba$); however, equations found by applying Kirchhoff's second rule to these additional loops will not be independent of the six equations found previously.

REVIEW CHECKLIST

- Calculate the terminal voltage, internal resistance, and current in a single-loop. (Section 18.1)

- Determine the equivalent resistance of a group of resistors in parallel, series, or series-parallel combination and find the current at any point in the circuit and the potential difference between any two points. (Sections 18.2 and 18.3)

- Apply Kirchhoff's rules to solve multiloop circuits; that is, find the currents at any point and the potential difference between any two points. (Section 18.4)

- Describe in qualitative terms the manner in which charge accumulates on a capacitor or current flow changes through a resistor with time in a series circuit with battery, capacitor, resistor, and switch. Use Equations (18.7) and (18.9) to calculate q at any time t and also to find the time t for which the ratio q/Q_0 will have a specified value. (Section 18.5)

SOLUTIONS TO SELECTED END-OF-CHAPTER PROBLEMS

3.   A lightbulb marked "75 W [at] 120 V" is screwed into a socket at one end of a long extension cord in which each of the two conductors has a resistance of 0.800 Ω. The other end of the extension cord is plugged into a 120-V outlet. Draw a circuit diagram, and find the actual power of the bulb in the circuit described.

Solution

If a lightbulb consumes 75.0 W of power when connected to a 120-V power source, the resistance of the filament in the bulb is given by $\mathcal{P} = (\Delta V)^2 / R$ as

$$R_{\text{filament}} = \frac{(\Delta V)^2}{\mathcal{P}} = \frac{(120 \text{ V})^2}{75.0 \text{ W}} = 192 \text{ } \Omega$$

The 192 Ω resistance of the filament is in series with each of the 0.800 Ω resistances of the conductors in the extension cord. This series combination of resistors is then connected to the 120-V source as shown in the circuit diagram given below.

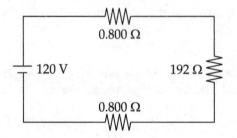

The equivalent resistance of this series combination is

$$R_{\text{eq}} = R_{\text{line}} + R_{\text{filament}} + R_{\text{line}} = 0.800 \text{ } \Omega + 192 \text{ } \Omega + 0.800 \text{ } \Omega$$

and the current which will flow through each resistance in the series combination is

$$I = \frac{\Delta V}{R_{\text{eq}}} = \frac{120 \text{ V}}{0.800 \text{ } \Omega + 192 \text{ } \Omega + 0.800 \text{ } \Omega} = 0.620 \text{ A}$$

The power actually delivered to the "75-W bulb" is then

$$\mathcal{P}_{\text{bulb}} = I^2 R_{\text{filament}} = (0.620 \text{ A})^2 (192 \text{ } \Omega) = 73.8 \text{ W}$$ ◊

9. Consider the circuit shown in Figure P18.9. Find (a) the current in the 20.0-Ω resistor and (b) the potential difference between points a and b.

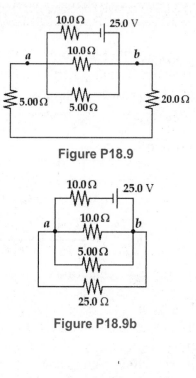

Figure P18.9

Solution

In Figure P18.9, the 20.0 Ω and the leftmost 5.00 Ω resistors are in series between points a and b. Replacing this series combination by its equivalent resistance of $R_{eq1} = 20.0\ \Omega + 5.00\ \Omega = 25.0\ \Omega$ reduces the circuit to that of Figure P18.9b. Then, it is seen that R_{eq1}, the lower 10.0 Ω resistor, and the remaining 5.00 Ω resistor are in parallel between points a and b. The equivalent resistance of this combination is

$$\frac{1}{R_{eq2}} = \frac{1}{10.0\ \Omega} + \frac{1}{5.00\ \Omega} + \frac{1}{25.0\ \Omega} = \frac{5 + 10 + 2}{50.0\ \Omega} \quad \text{or}$$

$$R_{eq2} = \frac{50.0\ \Omega}{17} = 2.94\ \Omega$$

Figure P18.9b

Replacing this parallel combination by its equivalent resistance yields the circuit shown in Figure P18.9c. Here, we see that R_{eq2} and the remaining 10.0 Ω resistor are connected in series across the battery. The equivalent resistance of this combination is $R_{total} = 10.0\ \Omega + 2.94\ \Omega = 12.94\ \Omega$, and the current the battery will supply to this series combination is

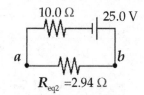

Figure P18.9c

$$I_{total} = \frac{\Delta V}{R_{total}} = \frac{25.0\ \text{V}}{12.94\ \Omega} = 1.93\ \text{A}$$

(b) The potential difference between points a and b is then

$$\Delta V_{ab} = I_{total} R_{eq2} = (1.93\ \text{A})(2.94\ \Omega) = 5.68\ \text{V} \qquad \Diamond$$

(a) From Figures P18.9b and P18.9, we see that the current through the lower branch of the circuit (i.e., the current through $R_{eq1} = 25.0\ \Omega$ in Figure P18.9b and also the current through each resistor of the series combination of the 20.0 Ω and leftmost 5.00 Ω resistors in Figure P18.9) will be

$$I_{R_{eq2}} = I_{20} = I_5 = \frac{\Delta V_{ab}}{R_{eq2}} = \frac{5.68\ \text{V}}{25.0\ \Omega} = 0.227\ \text{A} \qquad \Diamond$$

11.   The resistance between terminals *a* and *b* in Figure P18.11 is 75 Ω. If the resistors labeled *R* have the same value, determine *R*.

Solution

The figures given below, show the steps in simplifying the original circuit by using the rules for combining resistors in series and parallel.

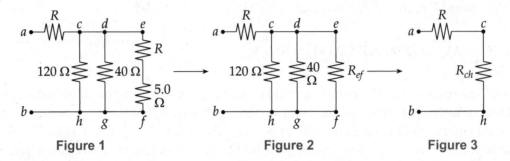

In Figure 1, observe that the two resistors between points *e* and *f* are in series. We replace this series combination by the single resistor $R_{ef} = R + 5.0$ Ω as shown in Figure 2. Then, from Figure 2, we observe that the three resistances R_{ef}, 40 Ω, and 120 Ω are connected in parallel. Recalling that $R_{ef} = R + 5.0$ Ω, the total resistance R_{ch} between points *c* and *h* is calculated using the rule for parallel resistors as

$$\frac{1}{R_{ch}} = \frac{1}{R+5.0\ \Omega} + \frac{1}{40\ \Omega} + \frac{1}{120\ \Omega} = \frac{4R+140\ \Omega}{(120\ \Omega)(R+5.0\ \Omega)} = \frac{R+35\ \Omega}{(30\ \Omega)(R+5.0\ \Omega)}$$

or $$R_{ch} = \frac{(30\ \Omega)(R+5.0\ \Omega)}{R+35\ \Omega}$$

Then, looking at Figure 3, observe that the total resistance between points *a* and *b* is $R_{ab} = R + R_{ch}$. Thus, if $R_{ab} = 75$ Ω, we must have

$$75\ \Omega = R + \frac{(30\ \Omega)(R+5.0\ \Omega)}{R+35\ \Omega} = \frac{R^2 +(65\ \Omega)R+150\ \Omega^2}{R+35\ \Omega}$$

which reduces to $R^2 - (10\ \Omega)R - 2\ 475\ \Omega^2 = 0$ or $(R - 55\ \Omega)(R + 45\ \Omega) = 0$

Only the positive solution is physically acceptable, so $R = 55$ Ω.   ◊

17. The ammeter shown in Figure P18.17 reads 2.00 A. Find I_1, I_2, and $\mathcal{E}$.

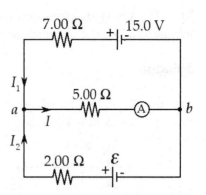

Solution

We are given that the current middle branch of the circuit is $I = 2.00$ A. Thus, the potential difference between points a and b in the circuit diagram at the right is

$$\Delta V_{ab} = (2.00 \text{ A})(5.00 \ \Omega) = 10.0 \text{ V}$$

Note that point a is at the higher potential since conventional current always flows toward points of lower potential and the current I flows from a toward b. Next, we start at point a and go counterclockwise around the upper loop recording changes in potential. It is essential to realize that as we go through the 7.00 Ω in the direction the conventional current I_1 flows, the change in potential is negative. Kirchhoff's loop rule gives

$$-10.0 \text{ V} + 15.0 \text{ V} - (7.00 \ \Omega)I_1 = 0 \qquad \text{or} \qquad I_1 = \frac{+5.00 \text{ V}}{7.00 \ \Omega} = 0.714 \text{ A} \qquad \Diamond$$

Then, applying Kirchhoff's junction rule at point a yields

$$I = I_1 + I_2 \qquad \text{or} \qquad I_2 = I - I_1 = 2.00 \text{ A} - 0.714 \text{ A} = 1.29 \text{ A} \qquad \Diamond$$

Finally, we start at point a and go clockwise around the loop forming the outer perimeter of the circuit. Note that as we pass through the 7.00 Ω resistor this time, we are going opposite to the flow of I_1 and the resulting change in potential will be positive. However, as we pass through the 2.00 Ω resistor, we will be going in the direction of the current I_2 and the change in potential will be negative. For this loop, Kirchhoff's loop rule gives

$$+(7.00 \ \Omega)I_1 - 15.0 \text{ V} + \mathcal{E} - (2.00 \ \Omega)I_2 = 0 \qquad \Diamond$$

so the emf of the battery in the lower branch is

$$\mathcal{E} = 15.0 \text{ V} + (2.00 \ \Omega)(1.29 \text{ A}) - (7.00 \ \Omega)(0.714 \text{ A}) = 12.6 \text{ V} \qquad \Diamond$$

23. Using Kirchhoff's rules, (a) find the current in each resistor shown in Figure P18.23 and (b) find the potential difference between points c and f.

Solution

(a) We assume the currents in the three branches of this circuit are directed as shown. If our assumed directions are correct, the calculated currents will have positive signs.

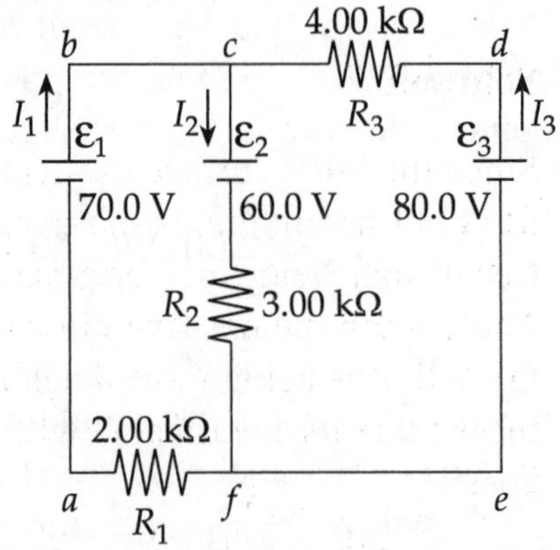

Apply Kirchhoff's junction rule at point c to obtain

$$I_2 = I_1 + I_3 \qquad (1)$$

Figure P18.23

Apply Kirchhoff's loop rule, going clockwise, around loop *abcfa*:

$$70.0 \text{ V} - 60.0 \text{ V} - (3\,000 \ \Omega)I_2 - (2\,000 \ \Omega)I_1 = 0$$

or $I_1 = 5.00 \times 10^{-3} \text{ A} - (1.50)I_2 \qquad (2)$

Apply Kirchhoff's loop rule, going counterclockwise, around loop *edcfe*:

$$80.0 \text{ V} - (4\,000 \ \Omega)I_3 - 60.0 \text{ V} - (3\,000 \ \Omega)I_2 = 0$$

or $I_3 = 5.00 \times 10^{-3} \text{ A} - (0.750)I_2 \qquad (3)$

Substituting Equations (2) and (3) into Equation (1) gives

$$I_2 = 5.00 \times 10^{-3} \text{ A} - (1.50)I_2 + 5.00 \times 10^{-3} \text{ A} - (0.750)I_2$$

which yields $I_2 = \dfrac{10.0 \times 10^{-3} \text{ A}}{3.25} = 3.08 \times 10^{-3} \text{ A} = 3.08 \text{ mA}$ ◊

Then, Equation (1) gives $I_1 = 5.00 \text{ mA} - (1.50)(3.08 \text{ mA}) = 0.380 \text{ mA}$ ◊

and Equation (3) gives $I_3 = 5.00 \text{ mA} - (0.750)(3.08 \text{ mA}) = 2.69 \text{ mA}$ ◊

(b) To determine the potential difference between points c and f, we start at point f and go up the center branch of the circuit to point c, recording all changes in potential that occur. This gives:

$$\Delta V_{cf} = V_c - V_f = (3.00 \text{ k}\Omega)I_2 + 60.0 \text{ V} = (3\,000 \ \Omega)(3.08 \times 10^{-3} \text{ A}) + 60.0 \text{ V}$$

or $\Delta V_{cf} = +69.2 \text{ V}$ (with point c at a higher potential than point f)   ◊

29. Find the potential difference across each resistor in Figure P18.29.

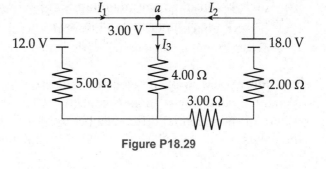

Figure P18.29

Solution

Assume that currents I_1, I_2, and I_3 are in the directions indicated by the arrows in the figure. Applying Kirchhoff's junction rule at junction a gives

$$I_3 = I_1 + I_2 \tag{1}$$

Using Kirchhoff's loop rule as you start at point a and go clockwise around the leftmost loop yields

$$-3.00 \text{ V} - (4.00 \text{ }\Omega)I_3 - (5.00 \text{ }\Omega)I_1 + 12.0 \text{ V} = 0$$

or $\qquad I_1 = 1.80 \text{ A} - 0.800 I_3 \tag{2}$

Starting at point a and going counterclockwise around the rightmost loop, Kirchhoff's loop rule gives

$$-3.00 \text{ V} - (4.00 \text{ }\Omega)I_3 - (3.00 \text{ }\Omega + 2.00 \text{ }\Omega)I_2 + 18.0 \text{ }V = 0$$

or $\qquad I_2 = 3.00 \text{ A} - 0.800 I_3 \tag{3}$

Substituting Equations (2) and (3) into Equation (1) gives

$$I_3 = 1.80 \text{ A} - 0.800 I_3 + 3.00 \text{ A} - 0.800 I_3 \quad \text{or} \quad I_3 = \frac{4.80 \text{ A}}{2.60} = 1.846 \text{ A}$$

Then, Equation (2) yields $\quad I_1 = 1.80 \text{ A} - 0.800(1.846 \text{ A}) = 0.323 \text{ A}$

and Equation (3) gives $\qquad I_2 = 3.00 \text{ A} - 0.800(1.846 \text{ A}) = 1.523 \text{ A}$

Therefore, the potential differences across the resistors are

$$\Delta V_2 = I_2(2.00 \text{ }\Omega) = 3.05 \text{ V} \qquad\qquad \Delta V_3 = I_2(3.00 \text{ }\Omega) = 4.57 \text{ V} \qquad\qquad \Diamond$$

$$\Delta V_4 = I_3(4.00 \text{ }\Omega) = 7.38 \text{ V} \qquad\qquad \Delta V_5 = I_1(5.00 \text{ }\Omega) = 1.62 \text{ V} \qquad\qquad \Diamond$$

34. A series combination of a 12-kΩ resistor and an unknown capacitor is connected to a 12-V battery. One second after the circuit is completed, the voltage across the capacitor is 10 V. Determine the capacitance of the capacitor.

Solution

If the circuit is completed by closing switch S at time $t = 0$, the charge stored on the capacitor at time t later is given by

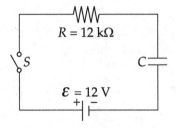

$$q = Q\left(1 - e^{-t/\tau}\right)$$

where $Q = C\mathcal{E}$ and the time constant is $\tau = RC$.

Thus, the potential difference across the capacitor at this time is

$$\Delta V = \frac{q}{C} = \frac{C\mathcal{E}\left(1 - e^{-t/\tau}\right)}{C} = \mathcal{E}\left(1 - e^{-t/\tau}\right)$$

If it is observed that $\Delta V = 10$ V at $t = 1.0$ s, then

$$10\text{ V} = \left(12\text{ V}\right)\left(1 - e^{-(1.0\text{ s})/\tau}\right) \qquad \text{or} \qquad e^{-(1.0\text{ s})/\tau} = 1 - \frac{10\text{ V}}{12\text{ V}} = \frac{1}{6}$$

Thus, $\quad e^{+(1.0\text{ s})/\tau} = 6 \quad$ giving $\quad \dfrac{1.0\text{ s}}{\tau} = \ln 6 \quad$ or $\quad \tau = \dfrac{1.0\text{ s}}{\ln 6}$

Since the time constant is $\tau = RC$, we find the unknown capacitance to be

$$C = \frac{\tau}{R} = \frac{1.0\text{ s}}{\left(12 \times 10^3\ \Omega\right)\ln 6} = 4.7 \times 10^{-5}\text{ F} = 47\ \mu\text{F}$$ ◊

---

39. A heating element in a stove is designed to dissipate 3 000 W when connected to 240 V. (a) Assuming that the resistance is constant, calculate the current in the heating element if it is connected to 120 V. (b) Calculate the power it dissipates at that voltage.

Solution

The power dissipated in the heating element is

$$\mathcal{P} = \left(\Delta V\right)I = \left(\Delta V\right)\left(\frac{\Delta V}{R}\right) = \frac{\left(\Delta V\right)^2}{R}$$

Thus, the resistance of the heating element in normal operating conditions is

$$R = \frac{\left(\Delta V\right)^2}{\mathcal{P}} = \frac{\left(240\text{ V}\right)^2}{3\,000\text{ W}} = 19.2\ \Omega$$

(a) If the resistance of the element is assumed to remain constant, the current when it is connected to a 120-V source will be

$$I' = \frac{\Delta V'}{R} = \frac{120 \text{ V}}{19.2 \text{ }\Omega} = 6.25 \text{ A}$$ ◊

(b) The power dissipated when connected to the 120-V source would be

$$\mathcal{P}' = (\Delta V')I' = (120 \text{ V})(6.25 \text{ A}) = 750 \text{ W}$$ ◊

Note that we could have also computed this power as

$$\mathcal{P}' = (I')^2 R = (6.25 \text{ A})^2 (19.2 \text{ }\Omega) = 750 \text{ W}$$ ◊

or as $$\mathcal{P}' = \frac{(\Delta V')^2}{R} = \frac{(120 \text{ V})^2}{19.2 \text{ }\Omega} = 750 \text{ W}$$ ◊

---

46. For the circuit shown in Figure P18.46, the voltmeter reads 6.0 V while the ammeter reads 3.0 mA. Find (a) the value of R, (b) the emf of the battery, and (c) the voltage across the 3.0-kΩ resistor. (d) What assumptions did you have to make to solve this problem?

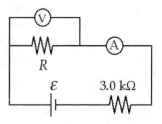

Figure P18.46

Solution

(a) The voltmeter gives a measure of the potential difference across the resistance R, $\Delta V = 6.0$ V. If we neglect any current passing through the voltmeter, the current measured by the ammeter is the same as the current through R, $I = 3.0$ mA $= 3.0 \times 10^{-3}$ A. Then, Ohm's law gives the resistance as

$$R = \frac{\Delta V}{I} = \frac{6.0 \text{ V}}{3.0 \times 10^{-3} \text{ }\Omega} = 2.0 \times 10^3 \text{ }\Omega = 2.0 \text{ k}\Omega$$ ◊

(b) Neglecting any current through the voltmeter, the current $I = 3.0 \times 10^{-3}$ A flows clockwise around the lower loop of the circuit (i.e., through R, through the ammeter, on through the 3.0 kΩ resistor, and back to the battery. Starting at the lower left corner of the circuit, going clockwise around this loop and neglecting any change in potential across the ammeter, Kirchhoff's loop rule gives $-RI - (3.0 \text{ k}\Omega)I + \mathcal{E} = 0$

or $\mathcal{E} = (R + 3.0 \text{ k}\Omega)I = (5.0 \times 10^3 \text{ }\Omega)(3.0 \times 10^{-3} \text{ A}) = 15 \text{ V}$ ◊

(c) The potential difference or voltage across the 3.0 kΩ resistor is

$$\Delta V_3 = (3.0 \text{ k}\Omega)I = (3.0 \times 10^3 \text{ }\Omega)(3.0 \times 10^{-3} \text{ A}) = 9.0 \text{ V}$$ ◊

(d) We have made three basic assumptions in this solution. First, we have assumed that the voltmeter and the ammeter are ideal devices. That is, the voltmeter has extremely high resistance and allows a negligible amount of current through it, while the ammeter has extremely low resistance and a negligible change in potential across it. Finally, we have assumed the battery is an ideal device, having a constant emf and negligible internal resistance.   ◊

---

55. The student engineer of a campus radio station wishes to verify the effectiveness of the lightning rod on the antenna mast (Fig. P18.55). The unknown resistance R_x is between points C and E. Point E is a "true ground," but is inaccessible for direct measurement, since the stratum in which it is located is several meters below the Earth's surface. Two identical rods are driven into the ground at A and B, introducing an unknown resistance R_y. The procedure for finding the unknown resistance R_x is as follows: Measure resistance R_1 between points A and B. Then connect A and B with a heavy conducting wire, and measure resistance R_2 between points A and C. (a) Derive a formula for R_x in terms of the observable resistances R_1 and R_2. (b) A satisfactory ground resistance would be $R_x < 2.0\ \Omega$. Is the grounding of the station adequate if measurements give $R_1 = 13\ \Omega$ and $R_2 = 6.0\ \Omega$?

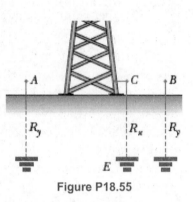

Figure P18.55

Solution

(a) For the first measurement (the resistance between points A and B), the equivalent circuit is as shown in Figure 1 at the right. Note that the lower ends of the resistances at A and B are connected by the Earth and that point C is not involved in this measurement. The measured resistance R_1 is the total resistance of a series combination

$$R_1 = R_y + R_y = 2R_y \qquad \text{so} \qquad R_y = R_1/2$$

Connecting points A and B makes a parallel combination of the two resistances labeled R_y as shown in Figure 2. This parallel combination has an equivalent resistance

$$\frac{1}{R_p} = \frac{1}{R_y} + \frac{1}{R_y} = \frac{2}{R_y} \qquad \text{or} \qquad R_p = \frac{R_y}{2} = \frac{R_1}{4}$$

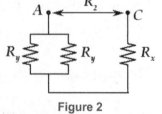

Figure 2

The measured R_2 is the total resistance of a series combination of R_p and R_x.

$$R_2 = R_p + R_x = \frac{R_1}{4} + R_x \qquad \text{so} \qquad R_x = R_2 - \frac{R_1}{4}$$

◊

(b) If $R_1 = 13 \ \Omega$ and $R_2 = 6.0 \ \Omega$, then $R_x = 6.0 \ \Omega - \dfrac{13 \ \Omega}{4} = 2.8 \ \Omega > 2.0 \ \Omega$ ◊

so the antenna is inadequately grounded.   ◊

---

60. The circuit in Figure P18.60 contains two resistors, $R_1 = 2.0 \ k\Omega$ and $R_2 = 3.0 \ k\Omega$, and two capacitors, $C_1 = 2.0 \ \mu F$ and $C_2 = 3.0 \ \mu F$, connected to a battery with emf $\mathcal{E} = 120$ V. If there are no charges on the capacitors before switch S is closed, determine the charges q_1 and q_2 on capacitors C_1 and C_2, respectively, as functions of time after the switch is closed. (*Hint*: First reconstruct the circuit so that it becomes a simple RC circuit containing a single resistor and single capacitor in series, connected to the battery, and then determine the total charge q stored in the circuit.]

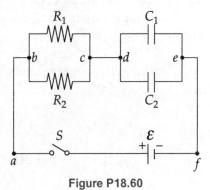

Figure P18.60

Solution

The equivalent resistance of the parallel combination of resistors between b and c is

$$\frac{1}{R_{bc}} = \frac{1}{2.0 \ k\Omega} + \frac{1}{3.0 \ k\Omega} = \frac{5.0}{6.0 \ k\Omega} \quad \text{or} \quad R_{bc} = \frac{6.0 \ k\Omega}{5.0} = 1.2 \ k\Omega$$

Also, the total capacitance between points d and e is $C_{de} = C_1 + C_2 = 5.0 \ \mu F$

so the time constant for this circuit is

$$\tau = R_{bc} C_{de} = \left(1.2 \times 10^3 \ \Omega\right)\left(5.0 \times 10^{-6} \ F\right) = 6.0 \times 10^{-3} \ s = 6.0 \ ms$$

Thus, if the switch is closed at $t = 0$, the total charge stored between points d and e at time $t > 0$ is

$$q_{de} = Q\left(1 - e^{-t/\tau}\right) = C_{de}\mathcal{E}\left(1 - e^{-t/6.0 \ ms}\right)$$

and the potential difference between d and e as a function of time is

$$\Delta V_{de} = \frac{q_{de}}{C_{de}} = \mathcal{E}\left(1 - e^{-t/6.0 \ ms}\right) = (120 \ V)\left(1 - e^{-t/6.0 \ ms}\right)$$

The charges on the capacitors C_1 and C_2 as functions of time are then

$$q_1 = C_1\left(\Delta V_{de}\right) = (2.0 \ \mu F)(120 \ V)\left(1 - e^{-t/6.0 \ ms}\right) = (240 \ \mu C)\left(1 - e^{-t/6.0 \ ms}\right)$$ ◊

and $$q_2 = C_2\left(\Delta V_{de}\right) = (3.0 \ \mu F)(120 \ V)\left(1 - e^{-t/6.0 \ ms}\right) = (360 \ \mu C)\left(1 - e^{-t/6.0 \ ms}\right)$$ ◊

63. What are the expected readings of the ammeter and voltmeter for the circuit in Figure P18.63?

Solution

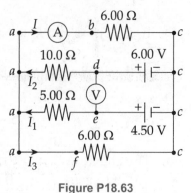

Figure P18.63

Note that all points labeled *a* in the circuit at the right are equivalent to each other and all the points labeled *c* are equivalent. The circuit could be redrawn with these merged into two single points *a* and *c* without altering the properties of the circuit.

To determine the voltmeter reading, we start at point *e* and follow the path *ecd* to point *d*, keeping track of all changes that occur in the potential. The result is

$$\Delta V_{de} = V_d - V_e = -4.50 \text{ V} + 6.00 \text{ V} = +1.50 \text{ V}$$

Thus, the voltmeter reads 1.50 V, with point *d* at the higher potential.   ◊

Starting on the currents, we apply Kirchhoff's loop rule, going clockwise around loop *abcfa* to obtain

$$-(6.00 \ \Omega)I + (6.00 \ \Omega)I_3 = 0 \quad \text{or} \quad I_3 = I \tag{1}$$

Next, apply Kirchhoff's loop rule, going clockwise around loop *abcda*:

$$-(6.00 \ \Omega)I + 6.00 \text{ V} - (10.0 \ \Omega)I_2 = 0 \quad \text{or} \quad I_2 = 0.600 \text{ A} - 0.600 I \tag{2}$$

Applying Kirchhoff's loop rule to the loop *abcea* gives

$$-(6.00 \ \Omega)I + 4.50 \text{ V} - (5.00 \ \Omega)I_1 = 0 \quad \text{or} \quad I_1 = 0.900 \text{ A} - 1.20 I \tag{3}$$

Finally, we apply Kirchhoff's junction rule at either point *a* or point *c* to obtain

$$I + I_3 = I_1 + I_2 \tag{4}$$

Substituting Equations (1), (2), and (3) into Equation (4) gives the current through the ammeter and hence the ammeter reading. The result of this substitution is

$$I + I = 0.900 \text{ A} - 1.20 I + 0.600 \text{ A} - 0.600 I$$

or $\quad 3.80 I = 1.50 \text{ A} \quad$ and $\quad I = \dfrac{1.50 \text{ A}}{3.80} = 0.395 \text{ A} = 395 \text{ mA}$ ◊

19

Magnetism

NOTES FROM SELECTED CHAPTER SECTIONS

19.1 Magnets

19.2 Earth's Magnetic Field
Like magnetic poles (north-north or south-south) repel each other, and unlike poles (north-south) attract each other. Magnetic poles cannot be isolated and always occur in north-south pairs. **Soft magnetic materials,** such as iron, are easily magnetized but also tend to lose their magnetism easily; **hard magnetic materials** such as cobalt and nickel are difficult to magnetize but tend to retain their magnetism.

The region of space surrounding any magnetic material can be characterized by a magnetic field; the magnetic field of the Earth can be pictured as having a pattern of magnetic field lines similar to the field lines surrounding a bar magnet. The magnetic pole of the Earth in the vicinity of the Earth's geographic North Pole corresponds to a magnetic south pole.

At any given location on the surface of the Earth a compass needle will indicate two angles related to the Earth's magnetic field:

Dip angle: the angle between the horizontal, relative to the surface of the Earth, and the direction of the magnetic field at that point.

Magnetic declination: the angle between true North (geographic North Pole) and magnetic North (indicated by a compass needle).

19.3 Magnetic Fields
The magnetic field is a vector quantity designated by the symbol $\vec{B}$ and has the SI unit tesla (T).

By convention the direction of the magnetic field vector is shown graphically as illustrated here. Magnetic field:

Lying in the plane of the page is shown by an arrow in the plane.

Directed out of the plane of the page is shown by the "tips" of arrows.

Directed into the plane of the page is shown by the "tails" of arrows.

A particle with charge q, moving with velocity $\vec{v}$ in a magnetic field $\vec{B}$, will experience a magnetic force $\vec{F}$ with magnitude $F = Bqv\sin\theta$. Properties of the magnetic force are:

- The magnetic force is proportional to the charge q and speed v of the particle.

- The magnitude of the magnetic force depends on the angle between the velocity vector of the particle and the magnetic field vector.

- When a charged particle moves in a direction parallel to the magnetic field vector, the magnetic force $\vec{F}$ on the charge is zero.

- The magnetic force acts in a direction perpendicular to both $\vec{v}$ and $\vec{B}$; that is, $\vec{F}$ is perpendicular to the plane formed by $\vec{v}$ and $\vec{B}$.

- The magnetic force on a positive charge is in the direction opposite to the force on a negative charge moving in the same direction.

- If the velocity vector makes an angle θ with the magnetic field, the magnitude of the magnetic force is proportional to $\sin\theta$.

The direction of the magnetic force on a **positive charge** can be determined by using **right-hand rule #1:**

1. With your thumb extended, point the fingers of your right hand in the direction of the velocity, $\vec{v}$.

2. Curl the fingers to point in the direction of the magnetic field, $\vec{B}$, moving through the smaller of the two possible angles.

3. Your thumb is now pointing in the direction of the magnetic force $\vec{F}$ exerted on a positive charge.

Another version of right-hand-rule #1 is described in Suggestions, Skills, and Strategies

19.4   Magnetic Force on a Current-Carrying Conductor

A magnetic force will be exerted on a current-carrying conductor when placed in a magnetic field. For a straight conductor, the magnitude of the force will be maximum when the conductor is perpendicular to the magnetic field and zero when the conductor is parallel to the magnetic field. *The net magnetic force on a closed current loop in a uniform magnetic field is zero.*

The direction of the force on a straight conductor can be determined by using **right-hand rule #1.**

In this case point the fingers on your right hand in the direction of conventional current (*I*) and curl your fingers to point in the direction of the magnetic field. Your thumb will now point in the direction of the magnetic force on the conductor.

Can you use the right-hand rule to determine the direction of the magnetic field that exerts a force as shown on the conductor at the right? (Ans. Into the plane of the page.)

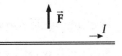

19.5   Torque on a Current Loop and Electric Motors

As stated above, zero net force is exerted on a closed current loop (coil) in a uniform magnetic field; however, there is a **net torque** exerted on a current loop in a magnetic field. The magnitude of the torque is proportional to the area of the loop, magnetic field strength, loop current, and the sine of the angle between the perpendicular to the plane of the loop and the direction of the magnetic field. The torque on the loop will be maximum when the field is parallel to the plane of the loop, and the torque will be zero when the field is perpendicular to the plane of the loop. *The torque will cause the loop to rotate so that the normal to the plane of the loop will turn toward the direction parallel to the magnetic field* (See figure on page 56.).

19.6   Motion of a Charged Particle in a Magnetic Field

A charged particle entering a uniform magnetic field with its velocity vector initially along a direction perpendicular to the field will move in a circular path in a plane perpendicular to the magnetic field. The magnetic force on the moving charge will be directed toward the center of the circular path and produce a centripetal acceleration. *The magnetic force changes the direction of the velocity vector but does not change its magnitude; the work done by the magnetic force on the particle is zero.*

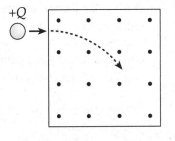

19.7   Magnetic Field of a Long, Straight Wire and Ampère's Law

A magnetic field will exist in the vicinity of a current-carrying wire. For a current in a long wire (where the length of the wire is much greater than the distance from the wire) the magnetic field is inversely proportional to the distance from the conductor. The direction of the magnetic field due to a current in a long straight wire is given by **right-hand rule #2:**

> **If the wire is grasped in the right hand with the thumb in the direction of the positive current, the fingers will wrap (or curl) in the direction of the circular magnetic field lines.**

As illustrated in the figure, the magnetic field lines are circular in a plane perpendicular to the length of the conductor.

At any point in space surrounding the conductor, the magnetic field vector, $\vec{B}$, is directed along the tangent to the circular field line through that point.

Ampère's circuital law states a method for calculating the magnetic field in the vicinity of a current-carrying conductor. *This technique is valid only for steady currents and can be easily used only in those cases where the current configuration has a high degree of symmetry.*

19.8   Magnetic Force Between Two Parallel Conductors

Parallel conductors carrying currents in the same direction attract each other; if the currents are in opposite directions, the conductors repel each other. *The forces on the two conductors are equal in magnitude regardless of the relative value of the two currents.*

The force between two parallel wires each carrying a current is used to define the ampere. If two long, parallel wires 1 m apart in a vacuum carry the same current and the force per unit length on each wire is 2×10^{-7} N/m, then the current is defined to be 1 A.

19.9   Magnetic Fields of Current Loops and Solenoids

The direction of the magnetic field at the center of a current loop is perpendicular to the plane of the loop and directed in the sense given by right-hand rule #2. For the situation illustrated in the figure, $\vec{B}$ is directed out of the page within the area of the loop and into the page outside the loop.

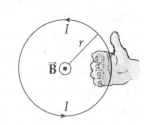

Within a solenoid, the magnetic field is parallel to the axis of the solenoid and pointing in a sense determined by applying right-hand rule #2 to one of the coils in the winding. *The magnetic field is uniform within the volume of the solenoid (except near the ends).*

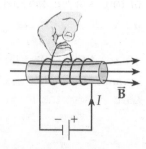

EQUATIONS AND CONCEPTS

The **magnitude of the magnetic force on a charge moving in a magnetic field** can have values ranging from zero to a maximum value, depending on the angle between the direction of the magnetic field and the direction of the velocity of the charge.

$$F = qvB\sin\theta \qquad (19.1)$$

θ is the smaller of the two angles between the directions of $\vec{v}$ and $\vec{B}$.

The **maximum value of the magnetic force** occurs when the charge is moving perpendicular to the direction of the magnetic field.

$$F_{max} = qvB \qquad (19.4)$$

The **magnitude of the magnetic field intensity** at some point in space is defined in terms of the magnetic force exerted on a moving positive electric charge at that point. Note in Equation (19.1) above that when $\theta = 0$, the force, $F = 0$. *Therefore, do not misinterpret Equation (19.2) to mean that B will be infinite when $\theta = 0$.*

$$B \equiv \frac{F}{qv\sin\theta} \qquad (19.2)$$

The **SI unit of magnetic field intensity** is the tesla (T).

$$[B] = T = Wb/m^2 = N/A \cdot m \qquad (19.3)$$

The **magnetic force exerted on a current-carrying conductor** placed in a magnetic field depends on the angle between the direction of the current in the conductor and the direction of the field. The magnetic force will be maximum when the conductor is directed perpendicular to the magnetic field. *Equation (19.6) applies in the case of a straight conductor.* What is the direction of force on the conductor in the figure?

$$F = BI\ell\sin\theta \qquad (19.6)$$

$$F_{max} = BI\ell \qquad (19.5)$$

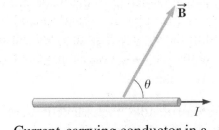

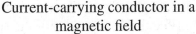

Current-carrying conductor in a magnetic field

The **magnitude of the net torque on a current loop** in an external magnetic field depends on the area of the loop, the angle between the direction of the magnetic field, and the direction of the normal (or perpendicular) to the plane of the loop, current, and magnitude of the magnetic field. For a multi-loop coil, the torque is proportional to the number of turns, N.

A **net torque** will be exerted on a current loop placed in an external magnetic field. In Equation (19.8), A has a magnitude equal to the area of the loop. *When the fingers of the right hand are curled around the loop in the direction of the current, the extended thumb points in the direction of* $\vec{A}$.

For a multi-loop coil, the torque is proportional to the number of turns, N.

- The **magnitude of the net torque** depends on the area of the loop, the angle between the direction of the magnetic field, and the direction of the normal (or perpendicular) to the plane of the loop.

- **Maximum magnitude of the torque** will occur when the magnetic field is parallel to the plane of the loop (i.e., $\theta = 90°$).

- The **direction of rotation of a current loop** in a magnetic field decreases the angle between the normal to the loop and the magnetic field; *the loop turns toward a position of minimum torque*. The loop shown in the figure will rotate counterclockwise as seen from above.

The **magnetic dipole moment of a current loop**, $\vec{\mu}$, can be used to express the torque on the loop when placed in a magnetic field. The direction of $\vec{\mu}$ is perpendicular to the plane of the loop, and θ is the angle between the directions of $\vec{\mu}$ and $\vec{B}$.

$$\tau = BIA \sin\theta \qquad (19.8)$$
(For a single loop)

$$\tau_{max} = BIA \qquad (19.7)$$

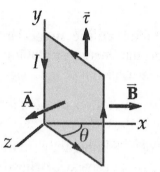

$$\tau = BIA \sin\theta \qquad (19.8)$$
(For a single loop)

$\theta =$ angle between the direction of $\vec{B}$ and the normal to the plane of the loop

$$\tau_{max} = BIA \qquad (19.7)$$

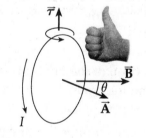

$$\mu = IAN$$

$$\tau = \mu B \sin\theta \qquad (19.9b)$$

Motion of a charged particle entering a uniform magnetic field with the velocity vector initially perpendicular to the field has the following characteristics:

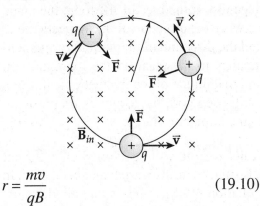

- The **path** of the particle will be circular and in a plane perpendicular to the direction of the field. The direction of rotation of the particle will be as determined by the right-hand rule.

- The **radius** of the circular path will be proportional to the linear momentum of the charged particle.

$$r = \frac{mv}{qB} \qquad (19.10)$$

- The **period of revolution** in the circular path described above is independent of the value of the radius.

$$T = \frac{2\pi r}{v} = \frac{2\pi m}{qB}$$

- A **helical path** will result if the initial velocity is not perpendicular to the magnetic field. The axis of the helix will be parallel to the magnetic field and the "pitch" (distance between adjacent coils) will depend on the component of velocity parallel to the field.

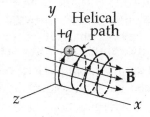

Equations for calculating the magnitude of the magnetic field due several current geometries with important applications are given below. The value of the permeability constant appearing in the equations is: $\mu_0 = 4\pi \times 10^{-7}$ T·m/A.

The magnitude of the magnetic field:

At a distance r from a **long straight conductor.**

$$B = \frac{\mu_0 I}{2\pi r} \qquad (19.11)$$

At the **center of a current loop** of N turns and radius R.

$$B = N\left(\frac{\mu_0 I}{2R}\right)$$

Inside a solenoid of n turns of conductor per unit length.

$$B = \mu_0 n I \qquad (19.16)$$
N = total turns
n = turns per meter

Ampère's circuital law can be used to find the magnetic field around certain simple current-carrying conductors. The product $(B_\parallel \Delta \ell)$ must be summed around a closed path containing the current.

$$\sum B_\parallel \Delta \ell = \mu_0 I \qquad (19.13)$$

$B_\parallel$ is the component of $\vec{B}$ tangent to a small displacement of length $\Delta \ell$.

The **magnitude of the force, per unit length, between parallel conductors** is proportional to the product of the two currents and inversely proportional to d, the distance between the two conductors. Parallel currents in the same direction attract each other, and parallel currents in opposite directions repel each other. *The forces on the two conductors will be equal in magnitude regardless of the relative magnitude of the two currents.*

$$\frac{F_1}{\ell} = \frac{\mu_0 I_1 I_2}{2\pi d} \qquad (19.14)$$

SUGGESTIONS, SKILLS, AND STRATEGIES

USING THE RIGHT-HAND RULE

Application of the right-hand-rule as illustrated below applies to *positive charges;* results are reversed in the case of negative charges. The right-hand rule is used to find the direction of force on a charged particle moving in a magnetic field. There are two versions of the right-hand rule; in order to avoid confusion, you should pick the version that suits you. In either version, note that the stated order of $\vec{v}$ and $\vec{B}$ is very important.

Version 1

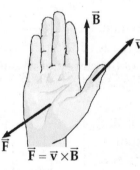

$$\vec{F} = \vec{v} \times \vec{B}$$

Version 2

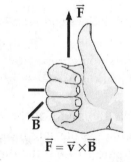

$$\vec{F} = \vec{v} \times \vec{B}$$

Hold your open right hand with your thumb pointing in the direction of $\vec{v}$ and your fingers pointing in the direction of $\vec{B}$. The force is now directed out of the palm of your hand. If the angle between $\vec{v}$ and $\vec{B}$ is zero, your thumb and fingers will be aligned and there is no force.

Orient your right hand so that your fingers point in the direction of $\vec{v}$ and then "curl" your fingers to point in the direction of $\vec{B}$. Your thumb now points in the direction of force. Since your fingers cannot bend more than $180°$, you may have to flip your hand upside down to accomplish the action described. If in a given situation, you do not need to curl your fingers because they already point along $\vec{v}$, the angle between $\vec{v}$ and $\vec{B}$ is zero and there is no force.

A different, but useful, right-hand rule applies in some specific situations. An example is when one of the quantities involves circulation (e.g., a current) and another a vector. The rule looks similar to version 2 described above; curl your fingers around in the direction of the circulation, and your thumb will point in the direction of the vector. This rule applies, for example, when finding the direction of the area vector for a current loop.

REVIEW CHECKLIST

- Use right-hand rule #1 and appropriate equations to determine the direction and the magnitude of the magnetic force exerted on a moving electric charge or current-carrying conductor in a region where there is a magnetic field. Practice the right-hand rule for situations in which different quantities (velocity, force, and magnetic field) represent the unknown direction. (Sections 19.3 and 19.4)

- Determine the magnitude of the torque exerted on a closed current loop in an external magnetic field and state the direction of rotation. (Section 19.5)

- Calculate the period and radius of the circular orbit of a charged particle moving in a uniform magnetic field. (Section 19.6)

- Calculate the magnitude and determine the direction (using right-hand rule #2) of the magnetic field for the following cases: (i) a distance r from a long, straight current-carrying conductor, (ii) at the center of a current loop of radius R, and (iii) at interior points of a solenoid with n turns per unit length. (Sections 19.7 and 19.9)

- Determine the magnetic force between parallel conductors. (Section 19.8)

SOLUTIONS TO SELECTED END-OF-CHAPTER PROBLEMS

3. Find the direction of the magnetic field acting on the positively charged particle moving in the various situations shown in Figure P19.3 if the direction of the magnetic force acting on it is as indicated.

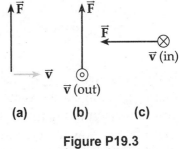

Figure P19.3

Solution

The magnetic force experienced by a moving charged particle is always perpendicular to both the velocity of the particle and the direction of the magnetic field. This means that the magnetic field, $\vec{B}$, must lie in the plane that is perpendicular to the force $\vec{F}$. However, without additional information, one cannot determine the exact orientation of $\vec{B}$ within that plane.

If we assume that the velocity, $\vec{v}$, of the particle is perpendicular to the magnetic field, right-hand rule #1 may be used to determine the direction of the field in each of the given situations.

Right-hand rule #1 is illustrated in the drawing at the right. Normally, this is used to find the direction of the magnetic force exerted on a positively charged particle, given the directions of the field and particle's velocity. Alternatively, we may use it to find the direction of the magnetic field, $\vec{B}$, from the known directions of the force and the velocity.

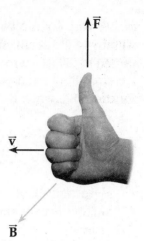

For each of the situations given in this problem, hold your right hand with the fingers extended in the given direction of $\vec{v}$ and the thumb pointing in the direction of the force exerted on the positive charge. Then, after the fingers have moved 90° as you are closing your hand, they will point in the direction of the magnetic field. This technique yields the following results:

(a) The field is directed into the page.   ◊

(b) $\vec{B}$ must be directed toward the right edge of the page.   ◊

(c) The direction of the field is toward the bottom of the page.   ◊

---

7.   What velocity would a proton need to circle Earth 1 000 km above the magnetic equator, where Earth's magnetic field is directed horizontally north and has a magnitude of 4.00×10^{-8} T?

Solution

The gravitational force acting on the proton will be negligible in comparison to the magnetic force. Thus, the magnetic force must supply the needed centripetal acceleration of the proton. To do this, the magnetic force must be directed downward, toward Earth, and have a magnitude given by

$$F = qvB\sin\theta = \frac{mv^2}{r}$$

Here, q is the charge of the proton, v is its speed, θ is the angle the velocity makes with magnetic field $\vec{B}$, m is the mass of the proton, and the radius of the proton's orbit is

$$r = R_E + 1\,000 \text{ km} = 6.38 \times 10^6 \text{ m} + 1\,000 \times 10^3 \text{ m} = 7.38 \times 10^6 \text{ m}$$

Since the proton moves parallel to the equator, and hence perpendicular to the south-to-north direction of the magnetic field, $\theta = 90.0°$ and the required speed is

$$v = \frac{qrB}{m} = \frac{\left(1.60 \times 10^{-19} \text{ C}\right)\left(7.38 \times 10^6 \text{ m}\right)\left(4.00 \times 10^{-8} \text{ T}\right)}{1.67 \times 10^{-27} \text{ kg}} = 2.83 \times 10^7 \text{ m/s} \qquad ◊$$

Note that the proton must move east to west around its orbit if the magnetic force is to be directed downward.   ◊

---

16. A wire having a mass per unit length of 0.500 g/cm carries a 2.00-A current horizontally to the south. What are the direction and magnitude of the minimum magnetic field needed to lift this wire vertically upward?

Solution

In order to lift the wire using a minimum magnetic field (and hence, minimum magnetic force), the magnetic force should be directed vertically upward and have a magnitude equal to the weight of the wire. If the wire has length L and carries current I through a magnetic field with magnitude B, the magnitude of the magnetic force acting on it is

$$F = BIL \sin\theta$$

where θ is the angle the direction of the current makes with the direction of the magnetic field. If this force has magnitude equal to the weight of the wire, then

$$BIL \sin\theta = mg \qquad \text{or} \qquad B = \frac{(m/L)g}{I \sin\theta}$$

Thus, for a minimum magnitude field, one should have $\theta = 90°$, or the field should be perpendicular to the current. The minimum acceptable magnitude for the field is

$$B_{min} = \left(\frac{m}{L}\right)\frac{g}{I \sin 90°} = \left[\left(0.500 \; \frac{g}{cm}\right)\left(\frac{1 \; kg}{10^3 \; g}\right)\left(\frac{10^2 \; cm}{1 \; m}\right)\right]\frac{(9.80 \; m/s^2)}{(2.00 \; A)(1)} = 0.245 \; T \qquad \Diamond$$

To determine the direction of this minimum magnitude magnetic field, use a variation of right-hand rule #1. With your right hand fully open, hold it so your extended fingers point in the direction of the current (horizontally southward) and the thumb points in the required direction of the force (vertically upward). Then, after moving the fingers 90° as you close your hand, the fingers point in the required direction of the field. You should find that this is horizontal and due east. $\qquad \Diamond$

19. A wire with a mass of 1.00 g/cm is placed on a horizontal surface with a coefficient of friction of 0.200. The wire carries a current of 1.50 A eastward and moves horizontally to the north. What are the magnitude and the direction of the *smallest* vertical magnetic field that enables the wire to move in this fashion?

Solution

The smallest vertical magnetic field which can move the wire in the desired manner will exert a horizontal force directed due northward (the direction we want the wire to move). Thus, $\vec{\mathbf{F}}$, $\vec{\mathbf{B}}$, and $\vec{\mathbf{I}}$ are all perpendicular to each other.

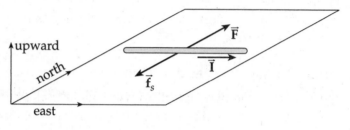

To determine whether $\vec{\mathbf{B}}$ must be vertically upward or downward, use the modified version of right-hand rule #1. Hold your right hand with the extended fingers pointing in the direction of the current in the wire (eastward) and the thumb pointing in the desired direction of the magnetic force (northward). As you close your hand, the fingers move toward the required direction of $\vec{\mathbf{B}}$. You should find that this is downward toward the ground.   ◊

To determine the required magnitude of the magnetic field, realize that if the magnetic force acting on the wire is horizontal, the normal force exerted on the wire by the horizontal surface is simply the weight, *mg*, of the wire. The maximum friction force retarding the motion of the wire is then

$$f_{s,\,\text{max}} = \mu_s n = \mu_s mg$$

When the magnetic field is just able to move the wire, the magnitude of the magnetic force equals the magnitude of the maximum friction force, or

$$F = f_{s,\,\text{max}} \quad \Rightarrow \quad BIL\sin 90.0° = \mu_s mg$$

With $m/L = 1.00$ g/cm $= 100$ g/m $= 0.100$ kg/m, this gives

$$B = \frac{\mu_s mg}{IL} = \frac{\mu_s (m/L)g}{I} = \frac{(0.200)(0.100 \text{ kg/m})(9.80 \text{ m/s}^2)}{1.50 \text{ A}} = 0.131 \text{ T} \qquad ◊$$

27. An eight-turn coil encloses an elliptical area having a major axis of 40.0 cm and a minor axis of 30.0 cm (Fig. P19.27). The coil lies in the plane of the page and has a 6.00-A current flowing clockwise around it. If the coil is in a uniform magnetic field of 2.00×10^{-4} T directed toward the left of the page, what is the magnitude of the torque on the coil? *Hint:* The area of an ellipse is $A = \pi ab$, where a and b are the semi-major and semi-minor axes, respectively, of the ellipse.

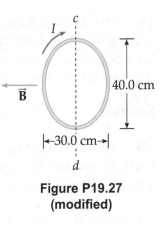

**Figure P19.27
(modified)**

Solution

The torque that a uniform magnetic field exerts on a flat (or planar) coil is given by

$$\tau = NBIA \sin\theta$$

where N is the number of turns of wire on the coil, B is the magnitude of the magnetic field, I is the current flowing in the coil, A is the area enclosed by the coil, and θ is the angle between the line perpendicular to the plane of the coil and the direction of the magnetic field. While the expression given above was derived for a rectangular coil in the textbook, it is valid for any planar coil. The area enclosed by the coil may be square, rectangular, circular, ellipsoidal, or any two-dimensional geometric shape.

In the situation shown in Figure P19.27, the magnetic field is parallel to the plane of the coil. Thus, the angle between the magnetic field and the line per pendicular to the plane of the coil is $\theta = 90.0°$. With semi-major and semi-minor axes of $a = 0.400$ m/2 and $b = 0.300$ m/2, the enclosed area of the coil is

$$A = \pi ab = \pi(0.200 \text{ m})(0.150 \text{ m}) = 0.030\,0\pi \text{ m}^2$$

and the magnitude of the torque exerted on the coil is

$$\tau = 8(2.00 \times 10^{-4} \text{ T})(6.00 \text{ A})(0.030\,0\pi \text{ m}^2)\sin 90.0° = 9.05 \times 10^{-4} \text{ N} \cdot \text{m} \qquad \lozenge$$

Note that, by right-hand rule #1, the magnetic force exerted on the left side of the coil is directed out of the page while the force exerted on the right side is into the page. Thus, an observer looking from c toward d along the dashed line shown in the figure would see the coil rotate counterclockwise.

37. A singly charged positive ion has a mass of 2.50×10^{-26} kg. After being accelerated through a potential difference of 250 V, the ion enters a magnetic field of 0.500 T, in a direction perpendicular to the field. Calculate the radius of the path of the ion in the field.

Solution

The singly charged $(q = 1e)$ ion starts from rest and accelerates through a 250-V potential difference before entering the magnetic field. Since electric fields are conservative force fields, the speed of the ion as it enters the field may be found using conservation of energy, $KE_f + PE_f = KE_i + PE_i$. The electrical potential energy of the ion at a location having potential V is $PE = qV$. Thus, we have

$$\frac{1}{2}mv^2 + qV_f = 0 + qV_i \qquad \text{or} \qquad v = \sqrt{\frac{2q|\Delta V|}{m}}$$

giving $v = \sqrt{\dfrac{2(1.60 \times 10^{-19}\ \text{C})(250\ \text{V})}{2.50 \times 10^{-26}\ \text{kg}}} = 5.66 \times 10^4$ m/s

When a charged particle moves perpendicularly to a magnetic field, the magnetic force exerted, always perpendicular to the motion, deflects the particle into a circular path and supplies the needed centripetal acceleration. That is,

$$F = qvB = m\frac{v^2}{r}$$

The radius of the path the particle follows is then $r = \dfrac{mv^2}{qvB} = \dfrac{mv}{qB}$. In this case, the radius is

$$r = \frac{(2.50 \times 10^{-26}\ \text{kg})(5.66 \times 10^4\ \text{m/s})}{(1.60 \times 10^{-19}\ \text{C})(0.500\ \text{T})} = 1.77 \times 10^{-2}\ \text{m} = 1.77\ \text{cm} \qquad \Diamond$$

46. In 1962, measurements of the magnetic field of a large tornado were made at the Geophysical Observatory in Tulsa, Oklahoma. If the magnitude of the tornado's field was $B = 1.50 \times 10^{-8}$ T pointing north when the tornado was 9.00 km east of the observatory, what current was carried up or down the funnel of the tornado? Model the vortex as a long, straight wire carrying a current.

Solution

Ionized gases, moving at fairly high speeds in a tornado, constitute electrical currents which can produce measurable magnetic fields. Here, we model the vortex as a long, straight, current-carrying conductor. The magnetic field lines produced by this current are circular, centered on the funnel of the tornado, and the magnitude of the field at distance r from the vortex is

$$B = \frac{\mu_0 I}{2\pi r}$$

Thus, if the measured field is $B = 1.50 \times 10^{-8}$ T at a distance $r = 9.00$ km $= 9.00 \times 10^3$ m from the tornado, the currents moving up or down the funnel must be

$$I = \frac{2\pi r B}{\mu_0} = \frac{2\pi \left(9.00 \times 10^3 \text{ m}\right)\left(1.50 \times 10^{-8} \text{ T}\right)}{4\pi \times 10^{-7} \text{ T} \cdot \text{m/A}} = 675 \text{ A} \qquad \Diamond$$

In the cited case, the tornado was located east of the observatory, or the observers were measuring the magnetic field on the *west side of the funnel*. To determine the direction of the current producing the observed northward directed field, imagine gripping the vertical funnel with your right hand so as your fingers curl around the vortex, they point northward (in the direction of the observed field) on the west side of the funnel. Then you should find that the thumb of your right hand is pointing downward, toward the ground. Therefore, the conventional current in the funnel was directed downward. $\qquad \Diamond$

Most likely, the current consisted of a mix of positive ions moving downward (in the direction of the conventional current) and free electrons (negative charges) moving upward, opposite the direction of the conventional current. Such movements of both types of charge would contribute to a northward directed magnetic field on the west side of the funnel.

51. A wire carries a 7.00-A current along the *x*-axis, and another wire carries a 6.00-A current along the *y*-axis, as shown in Figure P19.51. What is the magnetic field at point *P*, located at *x* = 4.00 m, *y* = 3.00 m?

Solution

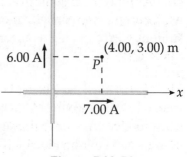

Let us call the wire along the *x*-axis wire 1 and that along the *y*-axis wire 2. Also, we shall consider a magnetic field at point *P* to be positive when directed out of the page toward the reader and negative if directed into the page.

Figure P19.51

The magnetic field at point *P* is the vector sum of the field generated by the current in wire 1 and the field generated by the current in wire 2. The magnitude of the field due to a current *I* in a long straight wire is given by

$$B = \frac{\mu_0 I}{2\pi r}$$

where *r* is the perpendicular distance from the observation point to the wire.

The direction of the field due to a long, straight conductor may be determined by imagining your right hand gripping the conductor with the thumb pointing along the conductor in the direction of the current. When this is done, the fingers circle around the conductor in the direction of the magnetic field lines produced by the current. Applying this technique to the wire along the *x*-axis shows that the contribution by wire 1 to the field at point *P* is directed out of the page and hence, is positive. Using the same method for the wire 2 indicates that its contribution to the field at *P* is directed into the page (negative direction).

Therefore, the resultant field at *P* is given by $\vec{\mathbf{B}} = \vec{\mathbf{B}}_1 + \vec{\mathbf{B}}_2$:

$$\left|\vec{\mathbf{B}}\right| = +\frac{\mu_0 I_1}{2\pi r_1} - \frac{\mu_0 I_2}{2\pi r_2} = \frac{4\pi \times 10^{-7}\ \text{T} \cdot \text{m/A}}{2\pi}\left(\frac{7.00\ \text{A}}{3.00\ \text{m}} - \frac{6.00\ \text{A}}{4.00\ \text{m}}\right) = +1.67 \times 10^{-7}\ \text{T}$$

or $\vec{\mathbf{B}} = 0.167\ \mu\text{T}$ directed out of the page   ◊

57. A wire with a weight per unit length of 0.080 N/m is suspended directly above a second wire. The top wire carries a current of 30.0 A and the bottom wire carries a current of 60.0 A. Find the distance of separation between the wires so that the top wire will be held in place by magnetic repulsion.

Solution

The sketch at the right shows the lower wire (wire 1) carrying a current I_1 directed out of the page. This current will produce a magnetic field whose field lines are circular and centered on this wire as shown. According to right-hand rule #2, the direction of this field is counterclockwise around wire 1, meaning that, at the location of the upper wire (wire 2), the field is directed to the left as shown. At the location of wire 2, distance r above wire 1, the magnitude of the field due to the current in wire 1 is $B_1 = \mu_0 I_1 / 2\pi r$.

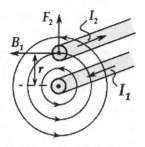

It is desired that wire 2 experience an upward force which will counteract the weight of this wire and cause it to float above wire 1. This force is the result of wire 2 carrying a current I_2 perpendicularly through the magnetic field generated by current in wire 1. Note that according to right-hand rule #1, the current in wire 2 must be directed into the page (opposite the direction of the current in wire 1) if the two wires are to repel each other.

The magnitude of the force experienced by wire 2, of length L_2, as it carries current I_2 through the field B_1 is

$$F_2 = B_1 I_2 L_2 = \left(\frac{\mu_0 I_1}{2\pi r}\right) I_2 L_2 = \frac{\mu_0 I_1 I_2 L_2}{2\pi r}$$

If this force is to support the weight of wire 2 and cause it to float above wire 1, it is necessary that $F_2 = m_2 g$, or

$$\frac{\mu_0 I_1 I_2 L_2}{2\pi r} = m_2 g \qquad \text{and} \qquad r = \frac{\mu_0 I_1 I_2}{2\pi (m_2 g / L_2)}$$

The distance separating the centers of the two wires should be

$$r = \frac{\left(4\pi \times 10^{-7} \text{ T} \cdot \text{m/A}\right)(60.0 \text{ A})(30.0 \text{ A})}{2\pi (0.080 \text{ N/m})} = 4.5 \times 10^{-3} \text{ m} = 4.5 \text{ mm} \qquad \lozenge$$

61. It is desired to construct a solenoid that will have a resistance of 5.00 Ω (at 20°C) and produce a magnetic field of 4.00×10^{-2} T at its center when it carries a current of 4.00 A. The solenoid is to be constructed from copper wire having a diameter of 0.500 mm. If the radius of the solenoid is to be 1.00 cm, determine (a) the number of turns of wire needed and (b) the length the solenoid should have.

Solution

From
$$R = \frac{\rho L_{\text{wire}}}{A} = \frac{\rho L_{\text{wire}}}{\pi d^2/4} = \frac{4\rho L_{\text{wire}}}{\pi d^2}$$

the length of copper wire with diameter $d = 0.500$ mm needed to provide the desired resistance is

$$L_{\text{wire}} = \frac{\pi d^2 R}{4\rho_{\text{Cu}}} = \frac{\pi (0.500 \times 10^{-3} \text{ m})^2 (5.00 \ \Omega)}{4(1.70 \times 10^{-8} \ \Omega \cdot \text{m})} = 57.7 \text{ m}$$

where the resistivity of copper was obtained from Table 17.1 in the textbook.

The number of turns this length of wire will make on the cylindrical form of the solenoid having a 1.00 cm radius is

$$N = \frac{L_{\text{wire}}}{2\pi r} = \frac{57.7 \text{ m}}{2\pi (1.00 \times 10^{-2} \text{ m})} = 918 \qquad \lozenge$$

The magnetic field inside a solenoid is given by $B = \mu_0 n I$, where $n = N/L_{\text{solenoid}}$ is the number of turns of wire per unit length on the solenoid. If it is desired to have $B = 4.00 \times 10^{-2}$ T when the current in the solenoid is $I = 4.00$ A, then

$$n = \frac{B}{\mu_0 I} = \frac{4.00 \times 10^{-2} \text{ T}}{(4\pi \times 10^{-7} \text{ T} \cdot \text{m/A})(4.00 \text{ A})} = 7.96 \times 10^3 \text{ turns/m}$$

and the required length for the solenoid is

$$L_{\text{solenoid}} = \frac{N}{n} = \frac{918 \text{ turns}}{7.96 \times 10^3 \text{ turns/m}} = 0.115 \text{ m} = 11.5 \text{ cm} \qquad \lozenge$$

69. Using an electromagnetic flowmeter (Fig. P19.69), a heart surgeon monitors the flow rate of blood through an artery. Electrodes *A* and *B* make contact with the outer surface of the blood vessel, which has interior diameter 3.00 mm. (a) For a magnetic field magnitude of 0.040 0 T, a potential difference of 160 μV appears between the electrodes. Calculate the speed of the blood. (b) Verify that electrode *A* is positive, as shown. Does the sign of the emf depend on whether the mobile ions in the blood are predominantly positively or negatively charged? Explain.

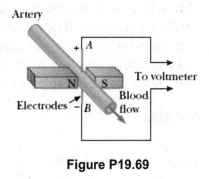

Figure P19.69

Solution

(a) The magnetic field inside the artery is directed from left to right (from the north (N) pole toward the south (S) pole of the magnet). According to right-hand rule #1, positive ions moving with the blood flow experience an upward magnetic force, deflecting them toward electrode *A*. Similarly, negative ions will experience a downward magnetic force and deflect toward electrode *B*. The accumulation of positive charge at *A* and negative charge at *B* creates an electric field directed from *A* toward *B*. The electric field exerts a downward force on positive ions and upward force on negative ions (opposite to the directions of the magnetic forces acting on the ions). At equilibrium, the magnitudes of the electric and magnetic forces are equal, leaving zero net force acting on the ions as they pass through the meter. Then,

$$qE = qvB \qquad \text{and the speed of the blood is} \qquad v = E/B$$

Since the magnitude of the electric field between the electrodes is $E = \Delta V/d$, where ΔV is the potential difference (i.e., the voltmeter reading) between *A* and *B*, and *d* is the distance between the electrodes (diameter of the artery), the speed of the blood is given by

$$v = \frac{\Delta V}{Bd} = \frac{160 \times 10^{-6} \text{ V}}{(0.040 \text{ 0 T})(3.00 \times 10^{-3} \text{ m})} = 1.33 \text{ m/s} \qquad \Diamond$$

(b) The polarity of the emf does not depend on the sign of the charge carriers. Positive charge carriers deflect upward to electrode *A*, making it positive relative to *B*. Likewise, negative charge carriers would deflect downward to electrode *B*, making it negative relative to *A*. These two situations are equivalent, yielding the same polarity emf (*A* at a higher potential than *B*). $\qquad \Diamond$

73. Protons having a kinetic energy of 5.00 MeV are moving in the positive *x*-direction and enter a magnetic field of 0.050 0 T in the *z*-direction, out of the plane of the page, and extending from $x = 0$ to $x = 1.00$ m as in Figure P19.73. (a) Calculate the *y*-component of the protons' momentum as they leave the magnetic field. (b) Find the angle α between the initial velocity vector of the proton beam and the velocity vector after the beam emerges from the field. *Hint:* Neglect relativistic effects and note that $1 \text{ eV} = 1.60 \times 10^{-19}$ J.

Solution

Note that we solve part (b) before solving part (a).

For a non-relativistic particle, the kinetic energy is $KE = \frac{1}{2}mv^2$, and the magnitude of the momentum may be expressed as $p = mv = \sqrt{2m(KE)}$.

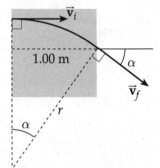

(b)   While the particle is in the magnetic field, it is moving perpendicular to the field and follows a circular arc. The magnetic force supplies the centripetal acceleration, so we determine the radius of the circular arc as follows:

$$qvB\sin 90° = m\frac{v^2}{r} \quad \text{and} \quad r = \frac{mv}{qB} = \frac{\sqrt{2m(KE)}}{qB}$$

Figure P19.73
(modified)

or $r = \dfrac{\sqrt{2\left(1.67 \times 10^{-27} \text{ kg}\right)\left(5.00 \times 10^{6} \text{ eV}\right)\left(1.60 \times 10^{-19} \text{ J/eV}\right)}}{\left(1.60 \times 10^{-19} \text{ C}\right)\left(0.050\ 0 \text{ T}\right)} = 6.46 \text{ m}$

Then, considering the sketch given above, observe that the angle between the directions of the initial and final velocities of the protons is

$$\alpha = \sin^{-1}\left(\frac{1.00 \text{ m}}{r}\right) = \sin^{-1}\left(\frac{1.00 \text{ m}}{6.46 \text{ m}}\right) = 8.90° \qquad ◊$$

(a)   Since the magnetic force is always perpendicular to the motion of the protons, it does no work on them, and they maintain constant speed while in the field. Thus, $v_f = v_i = \sqrt{2(KE)/m}$. As the protons emerge from the field, the *y*-component of their momentum is

$$\left(p_f\right)_y = m\left(v_f\right)_y = -mv_f \sin \alpha = -\sin \alpha \sqrt{2m(KE)}$$

or $\left(p_f\right)_y = -\sin(8.90°)\sqrt{2\left(1.67 \times 10^{-27} \text{ kg}\right)\left(5.00 \times 10^{6} \text{ eV}\right)\left(1.60 \times 10^{-19} \text{ J/eV}\right)}$

$$\left(p_f\right)_y = -8.00 \times 10^{-21} \text{ kg} \cdot \text{m/s} \qquad ◊$$

20

Induced Voltages and Inductance

NOTES FROM SELECTED CHAPTER SECTIONS

20.1 Induced EMF and Magnetic Flux

An electric current can be produced by a changing magnetic field. *An induced emf is produced in a secondary circuit by the changing magnetic field in a primary circuit.* The emf is induced by a change in a quantity called the magnetic flux rather than simply by a change in the magnetic field.

As described in Chapter 19, magnetic field lines can be drawn to illustrate the direction and relative strength of a magnetic field throughout a region of space. Consider a plane area, A, immersed in a magnetic field with field lines intersecting the area; the magnetic flux is the product of the magnetic field, the area, and the cosine of the angle between the direction of the magnetic field and the perpendicular (normal) to the area. Magnetic flux, Φ_B, is proportional to the number of magnetic field lines passing through the area and has SI units of webers (1 Wb = 1 T·m^2).

20.2 Faraday's Law of Induction

The emf induced in a circuit (or conducting loop) is proportional to the time rate of change of magnetic flux through the circuit; the resulting current is called an induced current. The polarity of the induced emf can be predicted by **Lenz's law:**

> **The polarity of an induced emf is such that it produces a current whose magnetic field opposes the change in magnetic flux through the loop. The induced current will be in the direction that will tend to maintain the original flux through the loop.**

20.3 Motional FMF

A motional emf (potential difference) will be maintained across a conductor moving in a magnetic field as long as the direction of motion through the field is not parallel to the field direction. If the motion is reversed, the polarity of the potential difference will also be reversed.

20.5 Generators

An **alternating current (AC) generator** consists of a wire loop rotated by some external means in a magnetic field. The ends of the loop are connected to **slip rings** that rotate with the loop; connections to the external circuit are made by stationary brushes in contact with the slip rings. The induced emf varies sinusoidally with time with a frequency of 60 Hz (in the US and Canada.)

In a **direct current (DC) generator** the contacts to the rotating loop are made by a split ring, or **commutator.** In this design, the output voltage always has the same polarity (always in the same direction) and varies in magnitude from zero to some maximum value.

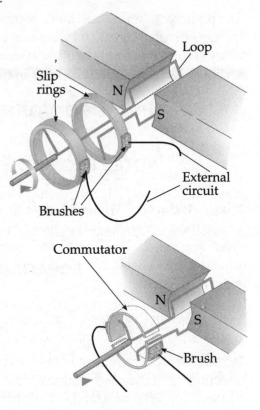

In the **operation of a motor,** a current is supplied to the loop by an external source of emf, and the magnetic torque on the current-carrying loop causes it to rotate, thereby doing mechanical work.

20.6 Self-Inductance

20.7 *RL* Circuits

When the switch is closed in a series circuit consisting of a resistor, a coil, and a battery, the current does not instantly go from zero to its maximum value. As the current increases in the coil, an induced emf is produced which opposes the emf of the battery. This effect, called **self-induction,** can be considered in the following steps:

- Current, initially at zero, increases through the coil resulting in an increasing magnetic flux through the coil.

- The changing flux induces an emf in the coil (Faraday's law).

- The induced emf is proportional to the rate at which the current is changing.

- The induced emf opposes the emf of the battery (Lenz's law).

- The net potential difference across the resistor is the emf of the battery minus the induced emf.

- The current continues to increase to a maximum value. *However, the rate at which the current increases becomes smaller.*

The property of a coil, solenoid, or other device (acting as an inductor) to limit the rate of change of current is called inductance (denoted by the symbol L) and has the SI unit of the henry (H). The inductance of a device (an inductor) depends on geometric factors (cross-sectional area, length, and number of turns of wire).

Remember, resistance (R) limits the current; inductance (L) limits the rate at which the current changes.

20.8   Energy Stored in a Magnetic Field

In an *RL* circuit, the rate at which energy is supplied by the battery equals the sum of the rate at which energy is dissipated in the resistor plus the rate at which energy is stored in the magnetic field of the inductor. *The energy stored in the magnetic field of an inductor is proportional to the square of the current in that inductor.*

EQUATIONS AND CONCEPTS

The **total magnetic flux** (Φ_B) through a plane area placed in a uniform magnetic field depends on the angle between the direction of the magnetic field and the direction perpendicular (normal) to the surface area. The maximum flux through the area occurs when the magnetic field is perpendicular to the surface area (i.e., $\vec{B}$ is along the direction of the normal and $\theta = 0°$). The unit of magnet flux is the weber (Wb). $1 \text{ Wb} = 1 \text{ T} \cdot \text{m}^2$

$$\Phi_B \equiv B_\perp A = BA \cos\theta \tag{20.1}$$

$$\Phi_{B,\max} = BA$$

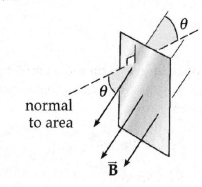

Faraday's law of induction states that the average emf induced in a circuit during a time interval Δt is proportional to the rate of change of magnetic flux through the circuit. *The minus sign is included to indicate the polarity of the induced emf, which can be found by use of Lenz's law.*

$$\mathcal{E} = -N \frac{\Delta \Phi_B}{\Delta t} \tag{20.2}$$

Lenz's law states that the polarity of the induced emf (and the direction of the associated current in a closed circuit) produces a current whose magnetic field opposes the change in the flux through the loop. *That is, the induced current tends to maintain the original flux through the circuit.* Note: If the circuit contains a source of emf (i.e., a battery) the current in the circuit may not be in the direction of the induced emf.

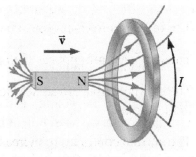

Lenz's law: The current induced in the loop produces a flux directed to the left.

A **motional emf** is induced in a conductor of length, ℓ, moving with speed, v, perpendicular to a magnetic field. *Remember the significance of the minus sign as required by Lenz's law.*

$$|\mathcal{E}| = \frac{\Delta \Phi_B}{\Delta t} = B\ell v \qquad (20.4)$$

An **induced current** will be present when a conductor moving in a magnetic field is part of a complete circuit of resistance, R. Equation (20.5) applies when the conductor moves perpendicular to the field. *Use the right-hand-rule and Lenz's law to confirm the direction of current as shown in the figure.*

$$I = \frac{|\mathcal{E}|}{R} = \frac{B\ell v}{R} \qquad (20.5)$$

Conductor moving to the right through a magnetic field directed into the page.

A **sinusoidally varying emf** is produced when a loop of wire with N turns and cross-sectional area A rotates with an angular velocity ω (measured in radians per second) in a magnetic field. For a given loop, the maximum value of the induced emf will be proportional to the angular velocity of the loop. Note the direction of the current shown in the figure.

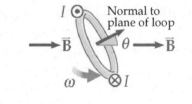

$$\mathcal{E} = NAB\omega \sin \omega t \qquad (20.7)$$

$$\mathcal{E}_{max} = NAB\omega \qquad (20.8)$$

Inductance (L) is characteristic of a conducting device (e.g., coil, solenoid, toroid, coaxial cable). The inductance of a given device, for example a coil, depends on its physical makeup: diameter, number of turns per unit length, type of material (core) on which the wire is wound, and other geometric parameters. The SI unit of inductance is the henry (H). A rate of change of current of 1 A/s in an inductor of 1 H will result in an induced voltage of 1 V.

$$1\,H = 1\,V \cdot s/A$$

An **inductor** is a circuit element.

Inductance is a measure of the degree to which an inductor opposes a change of current in a circuit.

The **inductance of a circuit element** can be expressed as the ratio of magnetic flux to current as shown in Equation (20.10) or as $L = -$ the ratio of induced emf to rate of change of current.

$$L = \frac{N\Phi_B}{I} \qquad (20.10)$$

$$L = -\frac{\mathcal{E}_L}{(\Delta I / \Delta t)}$$

The **inductance of a solenoid** can be calculated in terms of geometric quantities as shown in Equations (20.11a) and (20.11b)

$$L = \frac{\mu_0 N^2 A}{\ell} \qquad (20.11a)$$

$$L = \mu_0 n^2 V \qquad (20.11b)$$

$$n = N/\ell$$
$$V = A\ell = \text{volume of solenoid}$$

A **self-induced emf** ("back emf") will be generated in a coil when the current in the coil is changing. The emf will be proportional to the rate of change of the current and has a polarity that opposes the *change* occurring in the current. When the current increases as shown in the figure, the polarity of the self-induced emf will be as indicated by the dashed lines.

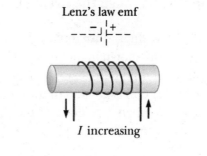

$$\mathcal{E}_L \equiv -L\frac{\Delta I}{\Delta t} \qquad (20.13)$$

A **series *RL* circuit** is shown in the figure at right. The circuit elements are a battery, resistor, inductor, and switch. *The inductor opposes any change in the current in the circuit.*

Current increases in an *RL* circuit (from $I=0$ to $I_{max} = \mathcal{E}/R$) in a characteristic fashion when the switch in the circuit shown above is moved to position 1. This is shown in the graph at right and described by the equation stated below the graph.

The **time constant** (τ) is the time required for the current to reach 63.2% of its maximum value. *The maximum current is achieved in a time that is long compared to the time constant, τ.*

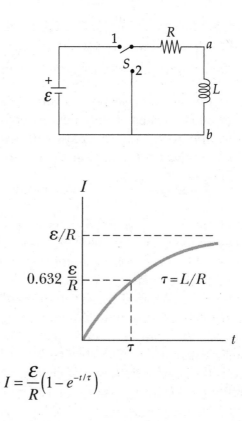

$$I = \frac{\mathcal{E}}{R}\left(1 - e^{-t/\tau}\right)$$

The **energy stored in the magnetic field** of an inductor is proportional to the square of the current.

$$PE_L = \tfrac{1}{2}LI^2 \qquad\qquad (20.15)$$

The **energy stored in the electric field** of a charged capacitor is proportional to the square of the potential difference between the plates.

$$PE_C = \tfrac{1}{2}C(\Delta V)^2$$

SUGGESTIONS, SKILLS, AND STRATEGIES

Remember, the induced current in a circuit will have a direction which tends to maintain the flux through the circuit. Carefully follow each step in the example below.

Consider two single-turn, concentric coils *lying in the plane of the paper* as shown in the figure. The outside coil is part of a circuit containing a resistor (R), a battery ($\mathcal{E}$), and switch (S). The inner coil is not part of the circuit. When the switch is moved from "**open**" to "**closed**" the direction of the induced current in the inner coil can be predicted by Lenz's law.

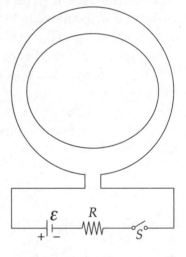

Consider the following steps:

(1)   When the switch is in the "**open**" position as shown, there will be no current in the circuit.

(2)   When the switch is moved to the "**closed**" position, there will be a clockwise current in the outside coil.

(3)   Magnetic field lines due to current in the outside coil will be directed into the page through the area enclosed by the coil. Use right-hand rule number 2 to confirm this. **Magnetic flux into the page will penetrate the entire area enclosed by the outside coil including the area of the inner coil.**

(4)   By Faraday's law, the increasing flux produces an induced emf (and current) in the inner coil.

(5)   Lenz's law requires that the induced current have a direction which will tend to maintain the initial flux condition (*which in this case was zero*). By applying right-hand number 2 to the inner coil, you should be able to determine that the direction of the induced current in the inner coil must be counterclockwise, contributing to a flux out of the page. Try this!

As a second example you should follow steps similar to those above to predict the direction of the induced current in the inner coil when the switch is moved from "**closed**" to "**open**."

REVIEW CHECKLIST

- Calculate the emf (or current) induced in a circuit when the magnetic flux through the circuit is changing in time. The variation in flux might be due to a change in: (i) the area of the circuit, (ii) the magnitude of the magnetic field, (iii) the direction of the magnetic field, or (iv) the orientation/location of the circuit in the magnetic field. Sections (20.1–20.2)

- Apply Lenz's law to determine the direction of an induced emf or current. You should also understand that Lenz's law is a consequence of the law of conservation of energy. Sections (20.2 and 20.4)

- Calculate the emf induced between the ends of a conducting bar as it moves through a region where there is a constant magnetic field (motional emf). Section (20.3)

- Define self-inductance, *L*, of a circuit in terms of appropriate circuit parameters. Calculate the total magnetic energy stored in a magnetic field if you are given the values of the inductance of the device with which the field is associated and the current in the circuit. Sections (20.6 and 20.8)

- Qualitatively describe the manner in which the instantaneous value of the current in an *RL* circuit changes while the current is either increasing or decreasing with time. Section (20.7)

SOLUTIONS TO SELECTED END-OF-CHAPTER PROBLEMS

5.   A long, straight wire lies in the plane of a circular coil with a radius of 0.010 m. The wire carries a current of 2.0 A and is placed along a diameter of the coil. (a) What is the net flux through the coil? (b) If the wire passes through the center of the coil and is perpendicular to the plane of the coil, find the net flux through the coil.

Solution

The magnetic field lines produced by a current in a long, straight wire are circular, centered on the wire, and in a plane perpendicular to the wire as shown in Figure (a). Note that if the current is toward the top of the page, the field lines go into the page on the right side of the wire and come out of the page on the left side of the wire.

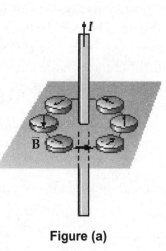

Figure (a)

(a)   If the long, straight wire is along the diameter of a flat
circular coil as shown in Figure (b), every magnetic
field line passing through the area of the coil directed
into the page on the right side of the long, straight
wire reemerges from the page through the area of the
coil on the left side of the wire. Thus, there are just
as many lines passing through the area of the coil in
one direction as there are passing through this area
in the opposite direction, and the net flux through the
area enclosed by the coil is **zero.**

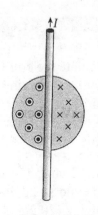

Figure (b)

(b)   If the wire passes through the center of the coil and is perpendicular to the plane of the
coil, then all of the field lines produced by the current in the long, straight wire lie in
planes parallel to the plane of the coil. Thus, none of these lines pass through the area
enclosed by the coil, and the flux through the coil is again **zero.**

---

11.   A wire loop of radius 0.30 m lies so that an external magnetic field of magnitude 0.30 T
is perpendicular to the loop. The field reverses its direction, and its magnitude changes to
0.20 T in 1.5 s. Find the magnitude of the average induced emf in the loop during this time.

Solution

The magnetic field is at all times
perpendicular to the plane of the
circular loop. Thus, the angle between
the direction of the field and the line
normal to the plane of the loop will be
either 0° or 180°.

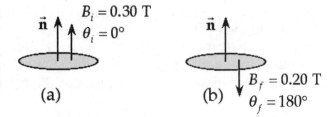

We choose the positive sense of the normal line to be in the original direction of the field
as shown in Figure (a) above, so $\theta_i = 0°$. After the field reverses direction, it is anti-parallel
to the chosen reference direction, and $\theta_f = 180°$ as in Figure (b). The change which occurs
in the flux through the circular loop as the field reverses direction and changes magnitude
during a time interval of $\Delta t = 1.5$ s is

$$\Delta\Phi_B = \Phi_{B,f} - \Phi_{B,i} = B_f A \cos\theta_f - B_i A \cos\theta_i = \left(B_f \cos\theta_f - B_i \cos\theta_i\right)A$$

$$= \left[(0.20\text{ T})\cos180° - (0.30\text{ T})\cos0°\right]\pi(0.30\text{ m})^2 = -0.14\text{ T}\cdot\text{m}^2 = -0.14\text{ Wb}$$

The magnitude of the average induced emf in the loop during this interval is then

$$\left|\mathcal{E}_{av}\right| = \frac{\left|\Delta\Phi_B\right|}{\Delta t} = \frac{0.14\text{ Wb}}{1.5\text{ s}} = 9.3\times10^{-2}\text{ V} = 93\text{ mV}$$

16. A circular coil enclosing an area of $100\ \text{cm}^2$ is made of 200 turns of copper wire. The wire making up the coil has resistance of $5.0\ \Omega$, and the ends of the wire are connected to form a closed circuit. Initially, a 1.1-T uniform magnetic field points perpendicularly upward through the plane of the coil. The direction of the field then reverses so that the final magnetic field has a magnitude of 1.1 T and points downward through the coil. If the time required for the field to reverse directions is 0.10 s, what average current flows through the coil during that time?

Solution

The magnitude of the average induced emf in a coil of fixed size is given by

$$\left|\mathcal{E}_{av}\right| = N\frac{\left|\Delta\Phi_B\right|}{\Delta t} = N\frac{\left|\Delta(B\cos\theta)\right|A}{\Delta t}$$

Here, N is the number of turns on the coil, B is the magnitude of the magnetic field, A is the area enclosed by the coil, and θ is the angle between the line normal to the plane of the coil and the direction of the magnetic field.

In the given situation, the field initially has a magnitude of 1.1 T and is directed parallel ($\theta = 0°$) to the normal line. After an interval of $\Delta t = 0.10$ s, the field has a magnitude of 1.1 T directed in the opposite ($\theta = 180°$) direction. Thus,

$$\left|\mathcal{E}_{av}\right| = (200)\frac{\left|(1.1\ \text{T})\cos180° - (1.1\ \text{T})\cos0°\right|(100\ \text{cm}^2)}{0.10\ \text{s}}\left(\frac{1\ \text{m}}{10^2\ \text{cm}}\right)^2 = 44\ \text{V}$$

and the average current in the coil during this interval is

$$I_{av} = \frac{\left|\mathcal{E}_{av}\right|}{R} = \frac{44\ \text{V}}{5.0\ \Omega} = 8.8\ \text{A}$$

◊

21. An automobile has a vertical radio antenna 1.20 m long. The automobile travels at 65.0 km/h on a horizontal road where Earth's magnetic field is $50.0\ \mu\text{T}$, directed toward the north and downwards at an angle of 65.0° below the horizontal. (a) Specify the direction the automobile should move in order to generate the maximum motional emf in the antenna, with the top of the antenna positive relative to the bottom. (b) Calculate the magnitude of this induced emf.

Solution

We want free electrons within the moving antenna to experience a downward force exerted on them by Earth's magnetic field. Since the magnetic force experienced by moving charges is perpendicular to the field exerting that force, only the horizontal component, B_h, of the Earth's field is effective in exerting a downward force on these electrons. The magnitude of this force is then

$$F = qvB_h\sin\theta \quad \text{where} \quad B_h = (50.0\ \mu\text{T})\cos65.0° = 21.1\ \mu\text{T}$$

is the northward directed horizontal component of Earth's field. Here, v is the magnitude of the antenna's velocity, and θ is the angle this velocity makes with B_h (that is, with the northward direction).

Electrons will migrate to the bottom of the antenna until an opposing force exerted by the developing electric field within the antenna has a magnitude equal to that of the magnetic force. When this occurs, $qE = qvB_h \sin\theta$ or $E = vB_h \sin\theta$, and the potential difference between the ends of an antenna of length ℓ is

$$\mathcal{E} = E\ell = v\ell B_h \sin\theta$$

(a)   If the emf discussed above is to have maximum magnitude, it is necessary that $\sin\theta = 1.00$ or $\theta = 90.0°$. Thus, the automobile must move along an east-west line. If the magnetic force on negative charges is to be downwards (and hence upward for positive charges), right-hand rule #1 shows that the automobile must move **eastward** along this line.   ◊

(b)   With $\theta = 90.0°$, the magnitude of the induced emf is

$$\mathcal{E} = v\ell B_h = \left[(65.0 \text{ km/h}) \left(\frac{0.278 \text{ m/s}}{1 \text{ km/h}} \right) \right] (1.20 \text{ m}) (21.1 \times 10^{-6} \text{ T})$$

or $\mathcal{E} = 4.58 \times 10^{-4} \text{ V} = 0.458 \text{ mV}$ ◊

---

25.   A bar magnet is positioned near a coil of wire as shown in Figure P20.25. What is the direction of the current in the resistor when the magnet is moved (a) to the left? (b) to the right?

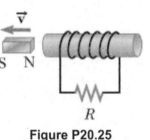

Figure P20.25

Solution

The movement of the magnet will produce a change in the flux through the turns of wire making up the coil. A changing flux through the coil induces an emf in the coil and produces a current in the wire and through the resistance R.

(a)   The induced current always flows in such a direction that the flux it produces opposes the change occurring in the flux due to the magnet. When the north end of the magnet in Figure P20.25 is moving to the left, away from the left end of the coil, the induced current must flow so the left end of the coil behaves as a south end of a magnet, attracting the departing north end of the bar magnet. This means that the magnetic field lines due to the current must enter the left end of the coil and emerge from the right end, pointing from left to right inside the coil. Imagine grasping the wire forming the leftmost turn on the coil with your right hand, such that your fingers curl around the wire and point from left to right (the desired direction of the field produced by the induced current) in the interior of the coil. Then, by right-hand rule 2, your thumb will point along the wire in the direction of the induced current. You should find that the current will be coming out of the page over the top of the coil, goes downward in the vertical segment of wire to the left of R, and flows from **left to right** through the resistance R.   ◊

(b) When the motion is toward the right, the north end of the magnet is approaching the left end of the coil. To oppose this, the left end of the coil must behave like the north end of a magnet, meaning that the magnetic field lines due to the induced current must emerge from the left end of the coil, loop around and reenter the right end. Thus, this field is directed from right to left through the interior of the coil. Imagine grasping the wire forming the leftmost turn on the coil with your right hand, so the fingers curl around the wire and point from right to left inside the coil. Then, by right-hand rule 2, you should find that the current flows from **right to left** through the resistance R, upward in the vertical wire to the left of R, and into the page at the top of the coil.   ◊

---

31. A rectangular coil with resistance R has N turns, each of length ℓ and width w as shown in Figure P20.31. The coil moves into a uniform magnetic field $\vec{\mathbf{B}}$ with constant velocity $\vec{\mathbf{v}}$. What are the magnitude and direction of the total magnetic force on the coil (a) as it enters the magnetic field, (b) as it moves within the field, and (c) as it leaves the field?

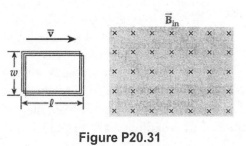

Figure P20.31

Solution

(a) During a time interval of duration Δt, after the right edge of the coil has entered the field and before the left edge reaches the field, the portion of the area enclosed by the coil which is inside the field (and has magnetic field lines passing through it) increases by an amount $\Delta A = w(v\Delta t)$. Thus, the flux through the area enclosed by each turn on the coil is directed into the page, and is increasing at a rate of $\Delta \Phi_B / \Delta t = B(\Delta A / \Delta t) = Bwv$. The magnitude of the emf induced in the coil is

$$|\mathcal{E}| = N \frac{|\Delta \Phi_B|}{\Delta t} = NBwv$$

The magnitude of the induced current is $I = |\mathcal{E}|/R = NBwv/R$, and it must flow counterclockwise around the coil in order to produce flux directed outward through the enclosed area (and thus oppose the increasing inward flux caused by the motion of the coil).

Note that the upper and lower edges of the coil (current carrying conductors, portions of which are in a magnetic field) experience magnetic forces that are equal in magnitude, oppositely directed, and cancel each other. However, each of the wires forming the right edge of the coil carries a current directed toward the top of the page through a magnetic field directed into the page. By right-hand rule 1, these wires experience magnetic forces directed toward the left. The wires forming the left edge of the coil are not yet in a magnetic field, and experience no magnetic forces. Thus, the net magnetic force on the coil is directed **toward the left** with a magnitude of

$$F_{\text{total}} = NF_{\text{single wire}} = NBIw = NB(NBwv/R)w = N^2B^2w^2v/R$$ ◊

(b) While the coil is entirely within the magnetic field, the flux through the area enclosed by the coil is constant. Hence, there is no induced emf or current in the coil, and the field exerts **zero force** on the coil.   ◊

(c) As the coil is emerging from the right edge of the field, an analysis similar to that in part (a) shows that the enclosed area with flux passing through it is decreasing at a rate $\Delta A/\Delta t = -wv$. Thus, the coil will experience an induced emf of magnitude $|\mathcal{E}| = N(|\Delta \Phi_B|/\Delta t) = NB(|\Delta A|/\Delta t) = NBwv$, and induced current of magnitude $I = |\mathcal{E}|/R = NBwv/R$. In this case, the induced current will flow clockwise around the loops of the coil so as to produce flux directed into the page through the enclosed area and oppose the decrease occurring in the inward flux due to the external field.

Again, the magnetic forces on the upper and lower edges of the coil will be equal in magnitude, opposite in direction, and will cancel. The right edge of the coil is now out of the field and experiences no magnetic force. Each of the N wires forming the left edge of the coil carry current I directed toward the top of the page through the magnetic field directed into the page. Right-hand rule 1 then shows that each of these wires experiences a magnetic force directed toward the left. The net magnetic force acting on the coil during this period is therefore directed **toward the left** with a total magnitude of

$$F_{\text{total}} = NF_{\text{single wire}} = NBIw = NB(NBwv/R)w = N^2 B^2 w^2 v/R \qquad ◊$$

---

37. In a model AC generator, a 500-turn rectangular coil 8.0 cm by 20 cm rotates at 120 rev/min in a uniform magnetic field of 0.60 T. (a) What is the maximum emf induced in the coil? (b) What is the instantaneous value of the emf in the coil at $t = (\pi/32)$ s ? Assume that the emf is zero at $t = 0$. (c) What is the smallest value of t for which the emf will have its maximum value?

Solution

When a coil rotates in a uniform magnetic field, the emf induced in that coil is given by $\mathcal{E} = NBA\omega \sin(\omega t)$. Here, N is the number of turns on the coil, B is the magnitude of the magnetic field, A is the area enclosed by the coil, and ω is the angular frequency of the coil.

(a) The angular frequency of the coil is

$$\omega = 120 \ \frac{\text{rev}}{\text{min}}\left(\frac{2\pi \ \text{rad}}{1 \ \text{rev}}\right)\left(\frac{1 \ \text{min}}{60 \ \text{s}}\right) = 4\pi \ \text{rad/s}$$

The maximum emf occurs at times when $\sin \omega t = 1.0$ and has a value of

$$\mathcal{E}_{\text{max}} = NBA\omega = (500)(0.60 \ \text{T})\big[(0.080 \ \text{m})(0.20 \ \text{m})\big](4\pi \ \text{rad/s}) = 60 \ \text{V} \qquad ◊$$

(b) At $t = (\pi/32)$ s, the instantaneous emf is

$$\mathcal{E} = \mathcal{E}_{max} \sin \omega t = (60 \text{ V})\sin\left[\left(4\pi \frac{\text{rad}}{\text{s}}\right)\left(\frac{\pi}{32} \text{ s}\right)\right] = 57 \text{ V}$$ ◊

Note that your calculator must be set to operate in radians mode (rather than degrees) to do the above calculation correctly.

(c) The first time $\sin \omega t$ (and hence $\mathcal{E} = \mathcal{E}_{max} \sin \omega t$) will achieve a maximum value is when

$$\omega t = \frac{\pi}{2} \text{ rad} \quad \text{or} \quad t = \frac{\pi \text{ rad}}{2\omega} = \frac{\pi \text{ rad}}{2(4\pi \text{ rad/s})} = 0.13 \text{ s}$$ ◊

---

44. An emf of 24.0 mV is induced in a 500-turn coil when the current is changing at a rate of 10.0 A/s. What is the magnetic flux through each turn of the coil at an instant when the current is 4.00 A?

Solution

The magnitude of the induced emf in a coil when the current changes at a rate $\Delta I/\Delta t$ is given by

$$|\mathcal{E}| = L\left(\frac{\Delta I}{\Delta t}\right) \tag{1}$$

where L is a physical property of the coil, called the self-inductance. The self-inductance of this coil is

$$L = \frac{|\mathcal{E}|}{(\Delta I/\Delta t)} = \frac{24.0 \times 10^{-3} \text{ V}}{10.0 \text{ A/s}} = 2.40 \times 10^{-3} \text{ } \Omega \cdot \text{s} = 2.40 \times 10^{-3} \text{ H} = 2.4 \text{ mH}$$

The magnitude of the induced emf in a coil may also be expressed as

$$|\mathcal{E}| = N\left(\frac{|\Delta\Phi_B|}{\Delta t}\right) \tag{2}$$

where N is the number of turns on the coil and Φ_B is the magnetic flux through each turn. Thus, equating equations (1) and (2), we obtain

$$N\left(\frac{|\Delta\Phi_B|}{\Delta t}\right) = L\left(\frac{\Delta I}{\Delta t}\right) \quad \text{or} \quad N|\Delta\Phi_B| = L(\Delta I)$$

Since N and L are constant for a given coil, it follows that at each instant in time, we must have $N\Phi_B = LI$. Thus, when the current in the coil is I, the magnetic flux through each turn on the coil must be $\Phi_B = LI/N$. For the given coil, the flux through each turn when the current is 4.00-A will be

$$\Phi_B = \frac{L \cdot I}{N} = \frac{(2.40 \times 10^{-3} \text{ H})(4.00 \text{ A})}{500} = 1.92 \times 10^{-5} \text{ T} \cdot \text{m}^2$$ ◊

---

47. A battery is connected in series with a 0.30 Ω resistor and an inductor as in Figure 20.27. The switch is closed at $t = 0$. The time constant of the circuit is 0.25 s and the maximum current in the circuit is 8.0 A. Find (a) the emf of the battery, (b) the inductance of the circuit, (c) the current in the circuit after one time constant has elapsed, and (d) the voltage across the resistor and the voltage across the inductor after one time constant has elapsed.

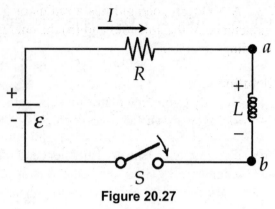

Figure 20.27

Solution

When switch S in the circuit of Figure 20.27 is closed at $t = 0$, a current given as a function of time by $i = I_{max}\left(1 - e^{-t/\tau}\right)$ begins to flow clockwise around the circuit. Here, $I_{max} = \mathcal{E}/R$ is the maximum current, reached only when $t \gg \tau$, and $\tau = L/R$ is known as the time constant of the circuit.

(a) From $I_{max} = \mathcal{E}/R$, the emf of the battery is seen to be

$$\mathcal{E} = I_{max}R = (8.0 \text{ A})(0.30 \text{ Ω}) = 2.4 \text{ V} \qquad \Diamond$$

(b) From $\tau = L/R$, the self-inductance in this circuit is found to be

$$L = \tau R = (0.25 \text{ s})(0.30 \text{ Ω}) = 7.5 \times 10^{-2} \text{ Ω·s} = 7.5 \times 10^{-2} \text{ H} = 75 \text{ mH} \qquad \Diamond$$

(c) At time $t = \tau$, the current in the circuit is

$$i = I_{max}\left(1 - e^{-\tau/\tau}\right) = I_{max}\left(1 - e^{-1}\right) = (8.0 \text{ A})(1 - 0.368) = 5.1 \text{ A} \qquad \Diamond$$

(d) Starting at point b and going clockwise around the circuit, Kirchhoff's loop rule gives $\mathcal{E} + \Delta v_R + \Delta v_L = 0$, where Δv_R and Δv_L are the changes in potential incurred across the resistor and across the inductor, respectively. Since we passed through the resistor in the direction of the current, $\Delta v_R = -IR$. Thus, at time $t = \tau$

$$\Delta v_R = -IR = -(5.1 \text{ A})(0.30 \text{ Ω}) = -1.5 \text{ V} \qquad \Diamond$$

and $\quad \Delta v_L = -\mathcal{E} - \Delta v_R = -2.4 \text{ V} - (-1.5 \text{ V}) = -0.90 \text{ V} \qquad \Diamond$

52. A 300-turn solenoid has a radius of 5.00 cm and a length of 20.0 cm. Find (a) the inductance of the solenoid and (b) the energy stored in the solenoid when the current in its windings is 0.500 A.

Solution

(a) The self-inductance L of an air-filled solenoid is a physical constant determined by the permeability of free space, μ_0, the number of turns on the solenoid, and the geometric factors ℓ and A. Here, ℓ is the length of the solenoid, and A is the cross-sectional area. The expression for the self-inductance in terms of these quantities is $L = \mu_0 N^2 A / \ell$. Thus, the self-inductance of the given solenoid is

$$L = \frac{\mu_0 N^2 A}{\ell} = \frac{\left(4\pi \times 10^{-7} \ \text{T} \cdot \text{m/A}\right)(300)^2 \left[\pi\left(5.00 \times 10^{-2} \ \text{m}\right)^2\right]}{20.0 \times 10^{-2} \ \text{m}}$$

or $L = 4.44 \times 10^{-3} \ \text{H} = 4.44 \ \text{mH}$ ◊

(b) The energy stored in the magnetic field of an inductor is

$$PE_L = \frac{1}{2} L I^2$$

where L is the self-inductance and I is the current flowing through the inductor. Therefore, when a current of $I = 0.500$ A exists in the windings of the solenoid described above, the stored energy is

$$PE_L = \frac{1}{2}\left(4.44 \times 10^{-3} \ \text{H}\right)(0.500 \ \text{A})^2 = 5.55 \times 10^{-4} \ \text{J} = 0.555 \times 10^{-3} \ \text{J} = 0.555 \ \text{mJ} \quad ◊$$

---

61. The bolt of lightning depicted in Figure P20.61 passes 200 m from a 100-turn coil oriented as shown. If the current in the lightning bolt falls from 6.02×10^6 A to zero in 10.5 μs, what is the average voltage induced in the coil? Assume that the distance to the center of the coil determines the average magnetic field at the coil's position. Treat the lightning bolt as a long, vertical wire.

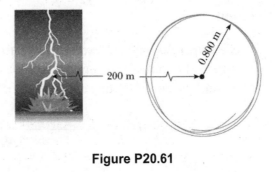

Figure P20.61

Solution

Treating the lightning bolt as a long vertical wire, the magnetic field lines produced by the current in the bolt will be horizontal circles centered on the bolt. Notice that this means the magnetic field is perpendicular to the plane of the 100-turn coil as it passes through the enclosed area of the coil. Also, because the diameter of the coil is small in comparison to the total distance to the lightning bolt, we will neglect the variation in magnitude of the magnetic field over the width of the coil.

When the current in the lightning bolt is I, the magnitude of the magnetic field produced at a distance of $r = 200$ m away is

$$B = \frac{\mu_0 I}{2\pi r} = \frac{\left(4\pi \times 10^{-7} \text{ T} \cdot \text{m/A}\right) I}{2\pi (200 \text{ m})} = \left(1.00 \times 10^{-9} \text{ T/A}\right) I$$

The average induced emf in the coil is then

$$\left|\mathcal{E}_{av}\right| = \frac{\left|\Delta \Phi_B\right|}{\Delta t} = \frac{\left|N(\Delta B)A\right|}{\Delta t} = \frac{100\left(1.00 \times 10^{-9} \text{ T/A}\right)\left|\Delta I\right| A}{\Delta t}$$

or

$$\left|\mathcal{E}_{av}\right| = \frac{100\left(1.00 \times 10^{-9} \text{ T/A}\right)\left|0 - 6.02 \times 10^6 \text{ A}\right| \pi (0.800 \text{ m})^2}{10.5 \times 10^{-6} \text{ s}} = 1.15 \times 10^5 \text{ V} = 115 \text{ kV} \quad \Diamond$$

---

65. In Figure P20.65, the rolling axle, 1.50 m long, is pushed along horizontal rails at a constant speed $v = 3.00$ m/s. A resistor $R = 0.400$ Ω is connected to the rails at points a and b, directly opposite each other. (The wheels make good electrical contact with the rails, so the axle, rails, and R form a closed-loop circuit. The only significant resistance in the circuit is R.) A uniform magnetic field $B = 0.800$ T is directed vertically downwards. (a) Find the induced current I in the resistor. (b) What horizontal force $\vec{F}$ is required to keep the axle rolling at constant speed? (c) Which end of the resistor, a or b, is at the higher electric potential? (d) After the axle rolls past the resistor, does the current in R reverse direction?

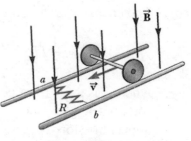

Figure P20.65

Solution

(a) As the rolling axle (of length $\ell = 1.50$ m) moves perpendicularly to the uniform magnetic field, an induced emf of magnitude $\left|\mathcal{E}\right| = B\ell v$ will exist between its ends. The current produced in the closed-loop circuit by this induced emf has magnitude

$$I = \frac{\left|\mathcal{E}\right|}{R} = \frac{B\ell v}{R} = \frac{(0.800 \text{ T})(1.50 \text{ m})(3.00 \text{ m/s})}{0.400 \text{ }\Omega} = 9.00 \text{ A} \quad \Diamond$$

(b) The induced current flowing through the axle will cause the magnetic field to exert a retarding force of magnitude $F_r = BI\ell$ on the axle. The direction of this force will be opposite to that of the velocity $\vec{v}$ so as to oppose the motion of the axle. If the axle is to continue moving at constant speed, an applied force in the direction of $\vec{v}$ and having magnitude $F_{app} = F_r$ must be exerted on the axle.

$$F_{app} = BI\ell = (0.800 \text{ T})(9.00 \text{ A})(1.50 \text{ m}) = 10.8 \text{ N} \quad \Diamond$$

(c) Using right-hand rule #1, observe that positive charges within the moving axle experience a magnetic force toward the rail containing point *b*, and negative charges experience a force directed toward the rail containing point *a*. Thus, the rail containing *b* will be positive relative to the other rail. Point *b* is then at a higher potential than *a*, and the conventional current goes from *b* to *a* through the resistor *R*.   ◊

(d) No. Both the velocity $\vec{v}$ of the rolling axle and the magnetic field $\vec{B}$ are unchanged. Thus, the polarity of the induced emf in the moving axle is unchanged, and the current continues to go from *b* to *a* through the resistor *R*.   ◊

21

Alternating Current Circuits
and Electromagnetic Waves

NOTES FROM SELECTED CHAPTER SECTIONS

21.1   Resistors in an AC Circuit

In an AC circuit both voltage and current vary sinsusoidally with time. If an alternating voltage is applied across a resistor, the current and voltage vary in the same manner, and both reach their maximum values at the same time. ***The current in the resistor will be in phase with the voltage.*** The voltage across a resistor is independent of the frequency and acts to limit the maximum value of the current.

The value of the rms current in an AC circuit is equal to that of a direct current which would deliver the same internal energy to a resistor as does the rms current.

21.2   Capacitors in an AC Circuit

When an alternating voltage is applied across a capacitor, the voltage reaches its maximum value one quarter of a cycle after the current reaches its maximum value. ***The voltage across a capacitor lags the current by 90°.***

A capacitor in an AC circuit acts to limit the current. The impeding effect of a capacitor is called **capacitative reactance, X_C.** The value of the capacitative reactance is inversely proportional to the frequency of the applied voltage.

21.3   Inductors in an AC Circuit

When an alternating voltage is applied across an inductor (coil), the voltage reaches its maximum value one quarter of a cycle before the current reaches its maximum value. ***The voltage across an inductor leads the current by 90°.*** The extent by which an inductor impedes the current in an AC circuit is called **inductive reactance, X_L.** This quantity is directly proportional to the frequency of the applied voltage.

21.4   The *RLC* Series Circuit

At any given time the current in a series *RLC* (resistor, inductor, and capacitor) circuit has the same amplitude and phase at all points in the circuit. The current through the individual elements reaches its maximum value at the same time. The voltages across the individual components (R, L, and C) do not reach their maximum values at the same time; they are out of phase with each other.

Examine the graphs below and confirm the following phase relationships:

The voltage across the resistor is in phase with the current.

The voltage across the capacitor lags the current by 90°.

The voltage across the inductor leads the current by 90°.

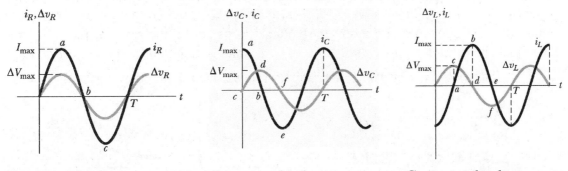

Current and voltage
vs. time for resistor

Current and voltage
vs. time for capacitor

Current and voltage
vs. time for inductor

The maximum applied voltage does not equal the sum of the individual maximum voltages:

$$\Delta V_{\max} \neq \Delta V_{R,\max} + \Delta V_{L,\max} + \Delta V_{C,\max}$$

The instantaneous applied voltage, however, does equal the sum of the instantaneous voltages across the individual components:

$$\Delta v = \Delta v_R + \Delta v_L + \Delta v_C$$

A **phasor diagram** illustrates the phase relationships among the voltages in an AC circuit. The maximum voltage across each circuit component is represented by a rotating vector called a phasor.

The **phasor diagram** is a very useful technique to use in the analysis of *RLC* circuits. In such a diagram, each of the quantities ΔV_R, ΔV_L, ΔV_C, and $I_{\max}$ is represented by a separate phasor (rotating vector).

A phasor diagram which describes the AC circuit of Figure (a) below is shown in Figure (b). Each phasor has a length which is proportional to the magnitude of the voltage or current which it represents and rotates counterclockwise about the common origin with a frequency which equals the frequency (f) of the alternating source.

The direction of the phasor which represents the current, shown as an open arrow in Figure (b), in the circuit is used as the *reference direction* to establish the correct phase differences among the phasors, which represent the voltage drops across the resistor, inductor, and capacitor. The *instantaneous values Δv_R, Δv_L, Δv_C, and i are given by the projection onto the vertical axis of the corresponding phasor.*

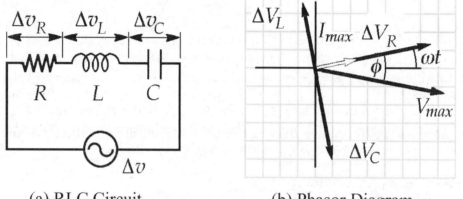

(a) RLC Circuit   (b) Phasor Diagram

Consider the phasor diagram in Figure (b), assume counterclockwise rotation, and note that:

(1) ΔV_L **leads the current by 90°** (ΔV_L reaches its maximum 90° before ΔV_R).

(2) ΔV_C **lags the current by 90°** (ΔV_C reaches maximum 90° later than ΔV_R).

(3) ΔV_R *greater than* ΔV_L; the maximum voltage across the resistor is **greater than** the maximum voltage across the inductor as seen by the lengths of the respective vectors.

(4) v_R *less than* v_L; the instantaneous voltage across the resistor is **less than** the instantaneous voltage across inductor as seen by the projections of the vectors onto the vertical axis.

Remember, the instantaneous values are the projections of the maximum values onto the vertical axis.

Also notice that as time increases and the phasors rotate counterclockwise, maintaining their constant relative phase, the voltage amplitudes (ΔV_R, ΔV_L, ΔV_C) will remain constant in magnitude; but the instantaneous values (Δv_R, Δv_L, and Δv_C) will vary sinusoidally with time. For the case shown in Figure (b), the phase angle ϕ is negative (this is because $X_C > X_L$ and therefore $\Delta V_C > \Delta V_L$); hence, the current in the circuit leads the applied voltage in phase.

The maximum voltage across an *RLC* circuit (ΔV_{max}) is found by adding ΔV_R, ΔV_L, and ΔV_C as vector quantities. In the figure at right, ΔV_{max} is shown as a rotating phasor. The **phase angle** (ϕ) indicates the degree by which the maximum voltage is out of phase with the current.

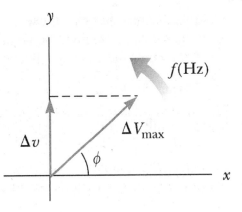

The **instantaneous voltage** (Δv) across the circuit is the component of ΔV_{max} along the *y*-axis. For example, when $\phi = 0$ or $180°$, $\Delta v = 0$; when $\phi = 90°$, $\Delta v = \Delta V_{max}$; and when $\phi = 270°$, $\Delta v = -\Delta V_{max}$.

The **impedance, Z,** is a parameter which represents the combined effect of R, X_C, and X_L in limiting current in an ac circuit. The value of Z depends on the values of R, L, C, and the frequency of the voltage source.

A **reactance (or impedance) triangle** showing R, X_L, X_C and Z can be drawn directly from the phasor diagram showing the voltages. This is possible because ΔV_R, ΔV_L, and ΔV_C are each proportional to the current for given values of R, X_L, and X_C.

21.5   Power in an AC Circuit
The power delivered by the generator in an AC circuit is converted to internal energy in the resistor. *There is no power loss in an ideal inductor or an ideal capacitor.* The power delivered to the circuit is determined by the power factor, $\cos \phi$.

21.6   Resonance in a Series *RLC* Circuit
As the frequency of the source (generator) increases, the resistance (R) in the circuit remains constant; inductive reactance (X_L) increases, and capacitive reactance (X_C) decreases. *At the resonance frequency, f_0 (when $X_L = X_C$), the voltage across the capacitor is equal to but 180 degrees out of phase (oppositely directed) with the voltage across the inductor. In this case, $\Delta V_L + \Delta V_C = 0$, and the impedance (Z) has its minimum value and is equal to R. At resonance the current in the circuit is maximum, limited only by the value of R.*

21.7   The Transformer
A transformer is a device designed to increase or decrease an AC voltage. Energy conservation requires a corresponding decrease or increase in current. In an ideal transformer, the power input to the primary equals the power output at the secondary (load). In practice, output power is less than input power; this is due in part to eddy currents induced in the transformer core. In its simplest form, a transformer consists of a primary coil of N_1 turns and a secondary coil of N_2 turns, both wound onto a common soft iron core. A step-up transformer has $N_2 > N_1$ and in a step-down transformer $N_1 > N_2$.

21.8   Maxwell's Predictions

21.9   Hertz's Confirmation of Maxwell's Predictions
Maxwell's equations are the fundamental laws governing the behavior of electric and magnetic fields. The theory Maxwell developed is based upon the following four pieces of information:

- Electric field lines originate on positive charges and terminate on negative charges. The electric field due to a point charge can be determined at a location by applying Coulomb's force law to a positive test charge placed at that location.

- Magnetic field lines always form closed loops; that is, they do not begin or end at any point.

- A varying magnetic field induces an emf and hence, an electric field. This is a statement of Faraday's law (Chapter 20).

- Magnetic fields are generated by moving charges (or currents), as summarized in Ampère's law (Chapter 19).

21.11 Properties of Electromagnetic Waves
Following is a summary of the properties of electromagnetic waves:

- **Electromagnetic waves are transverse waves** and travel through empty space with the speed of light, $c = 1/\sqrt{\epsilon_0 \mu_0}$.

- **Radiated waves exist as electric and magnetic fields which oscillate perpendicular to each other.** The plane in which the oscillations occur is perpendicular to the direction of wave propagation.

- **The ratio of $|\vec{E}|$ to $|\vec{B}|$ in empty space has a constant value,** the speed of light in vacuum.

- **Electromagnetic waves carry both energy and momentum.**

21.12 The Spectrum of Electromagnetic Waves
All electromagnetic waves are produced by accelerating charges. Several types of electromagnetic waves are characterized by a "typical" range of frequencies or wavelengths. The following are listed in order of increasing frequency.

- **Radio waves** (~10 km > λ > ~1 cm) are the result of electric charges accelerating through a conducting wire (antenna).

- **Microwaves** (~1 cm > λ > ~1 mm) are generated by electronic devices.

- **Infrared waves** (~1 mm > λ > ~700 nm) are produced by high temperature objects and molecules.

- **Visible light** (~700 nm > λ > ~400 nm) is produced by the rearrangement of electrons in atoms and molecules.

- **Ultraviolet (UV) light** (~400 nm > λ > ~1 nm) is an important component of radiation from the Sun.

- **X-rays** (~10 nm > λ > ~10^{-4} nm) are produced when high-energy electrons strike a target of high atomic number (e.g., metal or glass).

- **Gamma rays** (~10^{-1} nm > λ > ~10^{-5} nm) are emitted by radioactive nuclei.

EQUATIONS AND CONCEPTS

A **series alternating current circuit** with a sinusoidal source of emf is shown in the figure to the right. The rectangle ☐ used here represents the circuit element(s) which, in a particular case, may be a resistor (R), a capacitor (C), an inductor (L), or some combination of the above. The frequency of the current is determined by the AC source frequency, and the output voltage is sinusoidal as shown in Equation (21.1). *Keep in mind the different notations used to designate instantaneous, maximum, and rms values of voltage and current.*

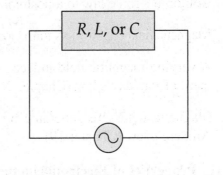

$$\Delta v = \Delta V_{max} \sin(2\pi f t) \tag{21.1}$$

Δv = instantaneous voltage
ΔV_{max} = maximum voltage

Root-mean-square (rms) values of current and voltage are those values to which measuring instruments usually respond.

$$I_{rms} = \frac{I_{max}}{\sqrt{2}} = 0.707\, I_{max} \tag{21.2}$$

$$\Delta V_{rms} = \frac{\Delta V_{max}}{\sqrt{2}} = 0.707\, \Delta V_{max} \tag{21.3}$$

Current in an *RLC* circuit can be limited by:

    Resistance (R) of the resistor

R is frequency independent

    Capacitive reactance (X_C) of the capacitor

$$X_C \equiv \frac{1}{2\pi f C} \tag{21.5}$$

    Inductive reactance (X_L) of the inductor

$$X_L \equiv 2\pi f L \tag{21.8}$$

The **series *RLC* circuit**, as shown at right, includes a resistor, inductor, capacitor, and a sinusoidally varying voltage source.

At any instant, the instantaneous current (i) is the same at all points in the circuit.

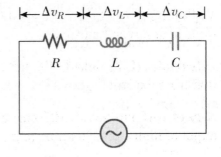

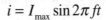

$$i = I_{max} \sin 2\pi f t$$

The **rms values of voltage and current in each circuit** element can be expressed by equations which have the form of Ohm's law. *Similar equations relate the maximum values of voltage and current.*

Resistor: $\Delta V_{R,\text{rms}} = I_{\text{rms}} R$ (21.4a)

Capacitor: $\Delta V_{C,\text{rms}} = I_{\text{rms}} X_C$ (21.6)

Inductor: $\Delta V_{L,\text{rms}} = I_{\text{rms}} X_L$ (21.9)

The **maximum (or rms) voltage** across the entire *RLC* circuit can be found in terms of:

the respective voltage values across the individual components

$$\Delta V_{\text{max}} = \sqrt{\Delta V_R^2 + (\Delta V_L - \Delta V_C)^2}$$ (21.10)

$$\Delta V_{\text{rms}} = \sqrt{\Delta V_{R,\text{rms}}^2 + (\Delta V_{L,\text{rms}} - \Delta V_{C,\text{rms}})^2}$$

or, the common circuit current and the values of resistance, inductive reactance, and capacitive reactance.

$$\Delta V_{\text{max}} = I_{\text{max}} \sqrt{R^2 + (X_L - X_C)^2}$$ (21.12)

$$\Delta V_{\text{rms}} = I_{\text{rms}} \sqrt{R^2 + (X_L - X_C)^2}$$

Impedance (*Z*) is a parameter of the circuit determined by *R*, X_L, and X_C and is frequency dependent. *The SI unit of impedance is the ohm.* The form of Equation (21.13) can be seen from the impedance triangle.

$$Z \equiv \sqrt{R^2 + (X_L - X_C)^2}$$ (21.13)

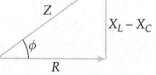

Impedance triangle

The **phase angle** (ϕ) is the degree by which the instantaneous current and instantaneous voltage are out-of-step. This parameter can be determined from the impedance triangle shown above and stated in Equation (21.15) or from a voltage triangle as expressed in Equation (21.11).

$$\tan \phi = \frac{X_L - X_C}{R}$$ (21.15)

$$\tan \phi = \frac{\Delta V_L - \Delta V_C}{\Delta V_R}$$ (21.11)

A **generalized form of Ohm's law** relates the maximum (or rms) voltage and maximum (or rms) current.

$$\Delta V_{\text{max}} = I_{\text{max}} Z$$ (21.14)

$$\Delta V_{\text{rms}} = I_{\text{rms}} Z$$

The **average power** delivered by a generator (source of emf) to an *RLC* series circuit is directly proportional to cos ϕ, the power factor of the circuit. *There is zero power loss in ideal inductors and capacitors; the average power delivered by the source is converted to internal energy in the resistor.*

$$\mathscr{P}_{av} = I_{rms}^2 R \qquad (21.16)$$

$$\mathscr{P}_{av} = I_{rms} \Delta V_{rms} \cos \phi \qquad (21.17)$$

$$\cos \phi = \text{power factor}$$

The **resonance frequency** (f_0) is that frequency for which $X_L = X_C$ (and $Z = R$). *At this frequency the current has its maximum value and is in phase with the applied voltage.*

$$f_0 = \frac{1}{2\pi\sqrt{LC}} \qquad (21.19)$$

A **transformer** consists of a primary coil of N_1 turns and a secondary coil of N_2 turns wound on a common core. *In a step-up transformer, N_2 is greater than N_1.*

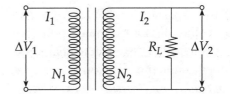

ΔV_1 = voltage across the primary
ΔV_2 = voltage across the secondary
I_1 = current in the primary
I_2 = current in the secondary

In **an ideal transformer,** the ratio of voltages is equal to the ratio of turns, and the ratio of currents is equal to the inverse of the ratio of turns. *Energy conservation requires a decrease in current to accompany an increase in voltage.*

$$\Delta V_2 = \frac{N_2}{N_1} \Delta V_1 \qquad (21.22)$$

$$I_1 \Delta V_1 = I_2 \Delta V_2 \qquad (21.23)$$

The **speed of an electromagnetic wave** is related to the permeability and permittivity of the medium through which it travels.

$$c = \frac{1}{\sqrt{\mu_0 \epsilon_0}} \qquad (21.24)$$

The **speed of light in a vacuum** equals the ratio of the magnitudes of the associated electric and magnetic fields.

$$\frac{E}{B} = c \qquad (21.26)$$

The **wave equation:** The product of frequency and wavelength of an electromagnetic wave propagating in vacuum is constant and equal to *c*.

$$c = f\lambda \qquad (21.31)$$

$$c = 2.997\,92 \times 10^8 \text{ m/s} \qquad (21.25)$$

Permeability constant of free space.

$$\mu_0 = 4\pi \times 10^{-7} \text{ T} \cdot \text{m/A}$$

Permittivity of free space.

$$\epsilon_0 = 8.854\ 19 \times 10^{-12} \text{ C}^2/\text{N} \cdot \text{m}^2$$

The **intensity (power per unit area) of electromagnetic waves** can be expressed in several alternate forms involving the maximum values of the electric and magnetic fields. The energy carried by an electromagnetic wave is shared equally by the electric and magnetic fields. *Do not confuse the symbol, I, representing intensity in Equations (21.27) and (21.28) with the same symbol used elsewhere for current.*

$$I = \frac{E_{max} B_{max}}{2\mu_0} \qquad (21.27)$$

$$I = \frac{E_{max}^2}{2\mu_0 c} = \frac{c}{2\mu_0} B_{max}^2 \qquad (21.28)$$

where

$$I = \frac{\mathcal{P}_{av}}{A} = \text{average power per unit area}$$

The **magnitude of the momentum** delivered to a surface by an electromagnetic wave depends on the fraction of energy absorbed. *Radiation pressure is exerted on a surface as a result of momentum transfer by an electromagnetic wave.*

$$p = \frac{U}{c} \text{ (complete absorption)} \qquad (21.29)$$

$$p = \frac{2U}{c} \text{ (complete reflection)} \qquad (21.30)$$

SUGGESTIONS, SKILLS, AND STRATEGIES

PROCEDURE FOR SOLVING ALTERNATING CURRENT PROBLEMS

- First calculate as many of the unknown quantities such as X_L and X_C as possible. Be careful to use correct units; when calculating X_C, express capacitance in farads (not microfarads).

- Apply the general equation $\Delta V = IZ$ to the portion of the circuit that is of interest. That is, if you want to know the voltage drop across the combination of an inductor and a resistor, the equation reduces to $\Delta V_{max} = I_{max}\sqrt{R^2 + X_L^2}$ or $\Delta V_{rms} = I_{rms}\sqrt{R^2 + X_L^2}$.

REVIEW CHECKLIST

- Given an *RLC* series circuit in which values of resistance, inductance, capacitance, and the characteristics of the generator (source of emf) are known, calculate: (i) the rms voltage drop across each component, (ii) the rms current in the circuit, (iii) the phase angle by which the current leads or lags the voltage, (iv) the power expended in the circuit, and (v) the resonance frequency of the circuit. (Sections 21.1–21.6)

- Understand the important function of transformers in the process of transmitting electrical power over large distances. Make calculations of primary to secondary voltage and current ratios for an ideal transformer. (Section 21.7)

- Be aware of the important pieces of information on which Maxwell based his theory of electromagnetic waves. Summarize the properties of electromagnetic waves. Describe the relative orientation of the magnetic field, electric field, and direction of propagation for a plane electromagnetic wave. (Sections 21.8–21.9)

- Place the various types of electromagnetic waves in the correct sequence in the electromagnetic spectrum. Be aware that all forms of radiation are produced by accelerated charges and know the basis of production particular to each of the wave types. (Section 21.11–21.12)

SOLUTIONS TO SELECTED END-OF-CHAPTER PROBLEMS

5. An audio amplifier, represented by the AC source and the resistor R in Figure P21.5, delivers alternating voltages at audio frequencies to the speaker. If the source puts out an alternating voltage of 15.0 V (rms), the resistance R is 8.20 Ω, and the speaker is equivalent to a resistance of 10.4 Ω, what is the time-averaged power delivered to the speaker?

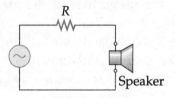

Figure P21.5

Solution

Observe that the resistance R and the resistance of the speaker R_{speaker} are in series in this purely resistive AC circuit. Thus, the total resistance of the circuit is

$$R_{\text{total}} = R + R_{\text{speaker}} = 8.20\ \Omega + 10.4\ \Omega = 18.6\ \Omega$$

The rms current in this single branch circuit is then

$$I_{\text{rms}} = \frac{\Delta V_{\text{rms, source}}}{R_{\text{total}}} = \frac{15.0\ \text{V}}{18.6\ \Omega} = 0.806\ \text{A}$$

and the time-averaged power delivered to the speaker is

$$\mathcal{P}_{\text{speaker}} = I_{\text{rms}}^2 R_{\text{speaker}} = (0.806\ \text{A})^2 (10.4\ \Omega) = 6.76\ \text{W} \qquad \lozenge$$

9. When a 4.0-μF capacitor is connected to a generator whose rms output is 30 V, the current in the circuit is observed to be 0.30 A. What is the frequency of the source?

Solution

When an AC voltage ΔV_C is applied to a capacitance C, the voltage and current are related by an Ohm's law-like expression, $\Delta V_C = IX_C$, where the capacitive reactance is given by

$$X_C = \frac{1}{2\pi fC} = \frac{1}{\omega C}$$

One may use either the maximum values for the voltage and current or the rms values for voltage and current since both $\Delta V_{C,max} = I_{max}X_C$ and $\Delta V_{C,rms} = I_{rms}X_C$ are valid expressions.

If $I_{rms} = 0.30$ A when $\Delta V_{C,\,rms} = 30$ V and $C = 4.0$ μF, the capacitive reactance of the capacitor is

$$X_C = \frac{\Delta V_{C,\,rms}}{I_{rms}} = \frac{30 \text{ V}}{0.30 \text{ A}} = 100 \text{ } \Omega$$

and the frequency of the AC voltage source must be

$$f = \frac{1}{2\pi X_C C} = \frac{1}{2\pi(100 \text{ } \Omega)(4.0 \times 10^{-6} \text{ F})} = 4.0 \times 10^2 \text{ Hz} \qquad \Diamond$$

---

17. Determine the maximum magnetic flux through an inductor connected to a standard outlet $(\Delta V_{rms} = 120 \text{ V}, f = 60.0 \text{ Hz})$.

Solution

The AC voltage across a pure inductance and the AC current through it is given by the Ohm's law-like expression $\Delta V_L = IX_L$, where the inductive reactance is $X_L = 2\pi fL = \omega L$. One may use either the maximum voltage and current or the rms values of voltage and current since both $\Delta V_{L,max} = I_{max}X_L$ and $\Delta V_{L,rms} = I_{rms}X_L$ are valid expressions.

As discussed in Section 20.6 in the textbook, the self-inductance of a coil may be expressed in terms of the total magnetic flux through the coil $\Phi_{B,total}$ as

$$L = \frac{\Phi_{B,total}}{I} = \frac{N\Phi_B}{I}$$

where N is the number of turns on the coil, and Φ_B is the flux through a single turn. Thus, the maximum total flux through the inductor is given by

$$\left(\Phi_{B,total}\right)_{max} = LI_{max} = \left(\frac{X_L}{2\pi f}\right)\left(\frac{\Delta V_{L,max}}{X_L}\right) = \frac{\sqrt{2}\left(\Delta V_{L,rms}\right)}{2\pi f}$$

where we have used the fact that $\Delta V_{max} = \sqrt{2}\left(\Delta V_{rms}\right)$ for a sinusoidal voltage.

Hence, if $\Delta V_{L,\mathrm{rms}} = 120$ V and $f = 60.0$ Hz, we find the maximum flux to be

$$\left(\Phi_{B,\mathrm{total}}\right)_{\max} = \frac{\sqrt{2}\,(120\ \mathrm{V})}{2\pi(60.0\ \mathrm{Hz})} = 0.450\ \mathrm{T\cdot m^2} \qquad \Diamond$$

---

21. A resistor $(R = 9.00 \times 10^2\ \Omega)$, a capacitor $(0.250\ \mu F)$, and an inductor $(L = 2.50\ \mathrm{H})$ are connected in series across a 2.40×10^2-Hz AC source for which $\Delta V_{\max} = 1.40 \times 10^2$ V. Calculate (a) the impedance of the circuit, (b) the maximum current delivered by the source, and (c) the phase angle between the current and voltage. (d) Is the current leading or lagging the voltage?

Solution

(a) The impedance of a series *RLC* circuit is given by $Z = \sqrt{R^2 + \left(X_L - X_C\right)^2}$, where R is the total resistance in the circuit, the inductive reactance is $X_L = 2\pi f L$ where L is the total self-inductance and f is the frequency of the power source, and $X_C = 1/2\pi f C$ is the capacitive reactance for a circuit containing total capacitance C. For the given circuit,

$$R = 900\ \Omega$$

$$X_L = 2\pi(240\ \mathrm{Hz})(2.50\ \mathrm{H}) = 3.77 \times 10^3\ \Omega$$

and $\quad X_C = \dfrac{1}{2\pi(240\ \mathrm{Hz})\left(0.250 \times 10^{-6}\ \mathrm{F}\right)} = 2.65 \times 10^3\ \Omega$

so the impedance is

$$Z = \sqrt{(900\ \Omega)^2 + \left(3.77 \times 10^3\ \Omega - 2.65 \times 10^3\ \Omega\right)^2} = 1.44 \times 10^3\ \Omega = 1.44\ \mathrm{k\Omega} \qquad \Diamond$$

(b) The maximum current delivered to this circuit by the source is

$$I_{\max} = \frac{\Delta V_{\mathrm{source,max}}}{Z} = \frac{140\ \mathrm{V}}{1.44 \times 10^3\ \Omega} = 9.72 \times 10^{-2}\ \mathrm{A} = 0.097\ 2\ \mathrm{A} \qquad \Diamond$$

(c) The difference in phase between the applied voltage and the current in this circuit is

$$\phi = \tan^{-1}\left(\frac{X_L - X_C}{R}\right) = \tan^{-1}\left(\frac{3.77 \times 10^3\ \Omega - 2.65 \times 10^3\ \Omega}{900\ \Omega}\right) = +51.2^\circ$$

(d) Since the inductive reactance exceeds the capacitive reactance, the phase angle ϕ is positive. This means that the applied **voltage leads the current** in this circuit. $\qquad \Diamond$

---

30. An AC source operating at 60 Hz with a maximum voltage of 170 V is connected in series with a resistor $(R = 1.2 \text{ k}\Omega)$ and a capacitor $(C = 2.5 \ \mu\text{F})$. (a) What is the maximum value of the current in the circuit? (b) What are the maximum values of the potential difference across the resistor and the capacitor? (c) When the current is zero, what are the magnitudes of the potential difference across the resistor, the capacitor, and the AC source? How much charge is on the capacitor at this instant? (d) When the current is at a maximum, what are the magnitudes of the potential differences across the resistor, the capacitor, and the AC source? How much charge is on the capacitor at this instant?

Solution

(a) The capacitive reactance for this circuit is

$$X_C = \frac{1}{2\pi fC} = \frac{1}{2\pi(60 \text{ Hz})(2.5 \times 10^{-6} \text{ F})} = 1.1 \times 10^3 \ \Omega = 1.1 \text{ k}\Omega$$

Therefore, the impedance of this *RC* circuit (with zero inductive reactance) is

$$Z = \sqrt{R^2 + X_C^2} = \sqrt{(1.2 \times 10^3 \ \Omega)^2 + (1.1 \times 10^3 \ \Omega)^2} = 1.6 \times 10^3 \ \Omega = 1.6 \text{ k}\Omega$$

and the maximum value of the current in the circuit is

$$I_{max} = \frac{\Delta V_{source,max}}{Z} = \frac{170 \text{ V}}{1.6 \times 10^3 \ \Omega} = 0.11 \text{ A} \qquad \diamond$$

(b) The maximum values of the voltages across the resistor and the capacitor are

$$\Delta V_{R,max} = RI_{max} = (1.2 \times 10^3 \ \Omega)(0.11 \text{ A}) = 1.3 \times 10^2 \text{ V} \qquad \diamond$$

and $\quad \Delta V_{C,max} = X_C I_{max} = (1.1 \times 10^3 \ \Omega)(0.11 \text{ A}) = 1.2 \times 10^2 \text{ V} \qquad \diamond$

(c)   When the instantaneous current is $i = 0$, the instantaneous voltage across the resistor is $\Delta v_R = iR = 0$. $\qquad \diamond$

The voltage across a capacitor is $90°$ or a quarter cycle out of phase with the current. Thus, when the instantaneous current is zero, the magnitude of the instantaneous voltage across the capacitor is a maximum, or $|\Delta v_C| = \Delta V_{C,max} = 1.2 \times 10^2 \text{ V}$. $\qquad \diamond$

The stored charge is then $q = C|\Delta v_C| = (2.5 \ \mu\text{F})(1.2 \times 10^2 \text{ V}) = 3.0 \times 10^2 \ \mu\text{C}$. $\qquad \diamond$

Applying Kirchhoff's loop rule at any instant to the single loop of this series circuit gives $\Delta v_{source} + \Delta v_R + \Delta v_C = 0$. Thus, at the instant when $i = 0$, $\Delta v_{source} + 0 - 1.2 \times 10^2 \text{ V} = 0$

and $|\Delta v_{source}| = 1.2 \times 10^2 \text{ V}$. $\qquad \diamond$

(d) When the instantaneous current is a maximum, the instantaneous voltage across the resistor ($\Delta v_R = iR$) will be a maximum, while the instantaneous voltage across the capacitor (a quarter cycle out of phase with the current) will be zero.

Thus, at this instant, we have $\left| \Delta v_R \right| = I_{max} R = \Delta V_{R,max} = 1.3 \times 10^2$ V, ◊

$$\left| \Delta v_C \right| = 0 \quad \text{and} \quad q = C \left| \Delta v_C \right| = 0.$$ ◊

Applying Kirchhoff's loop rule to the circuit at this instant gives

$$\Delta v_{source} - 1.3 \times 10^2 \text{ V} + 0 = 0, \text{ or } \left| \Delta v_{source} \right| = 1.3 \times 10^2 \text{ V}.$$ ◊

---

35. An inductor and a resistor are connected in series. When connected to a 60-Hz, 90-V (rms) source, the voltage drop across the resistor is found to be 50 V (rms), and the power delivered to the circuit is 14 W. Find (a) the value of the resistance and (b) the value of the inductance.

Solution

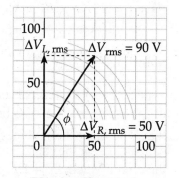

(a) The average power delivered to an AC circuit is

$$\mathcal{P}_{av} = I_{rms} \left(\Delta V_{rms} \right) \cos \phi$$

From the phasor diagram at the right, it is clear that the rms voltage across the resistor is

$$\Delta V_{R,rms} = \left(\Delta V_{rms} \right) \cos \phi$$

Phasor diagram for an *RL* circuit

Thus, the power can be written as

$$\mathcal{P}_{av} = I_{rms} \left(\Delta V_{R,rms} \right)$$

and the rms current in the circuit is $\quad I_{rms} = \dfrac{\mathcal{P}_{av}}{\Delta V_{R,rms}} = \dfrac{14 \text{ W}}{50 \text{ V}} = 0.28 \text{ A}$

Thus, the resistance of the resistor in the circuit is given by

$$R = \frac{\Delta V_{R,rms}}{I_{rms}} = \frac{50 \text{ V}}{0.28 \text{ A}} = 1.8 \times 10^2 \ \Omega$$ ◊

(b)   Observe from the phasor diagram that $\Delta V_{rms} = \sqrt{\left(\Delta V_{R,rms}\right)^2 + \left(\Delta V_{L,rms}\right)^2}$

The voltage across the inductor is then $\Delta V_{L,rms} = \sqrt{\left(\Delta V_{rms}\right)^2 - \left(\Delta V_{R,rms}\right)^2}$

or
$$\Delta V_{L,rms} = \sqrt{\left(90 \text{ V}\right)^2 - \left(50 \text{ V}\right)^2} = 75 \text{ V}$$

The inductive reactance,
$$X_L = \Delta V_{L,rms}/I_{rms}$$

is then
$$X_L = 75 \text{ V}/0.28 \text{ A} = 2.7 \times 10^2 \ \Omega$$

And the self-inductance is
$$L = \frac{X_L}{\omega} = \frac{2.7 \times 10^2 \ \Omega}{2\pi\left(60 \text{ Hz}\right)} = 0.71 \text{ H}$$ ◊

---

39.   The AM band extends from approximately 500 kHz to 1 600 kHz. If a 2.0-μH inductor is used in a tuning circuit for a radio, what are the extremes that a capacitor must reach in order to cover the complete band of frequencies?

Solution

A radio is tuned to a particular station when the resonance frequency of the tuning circuit in the radio matches the broadcast frequency of the desired station. Commonly, the tuning circuit is a series *RLC* circuit in which the antenna serves as the power source, and the capacitance of the capacitor is varied to adjust the resonance frequency. Resonance occurs in the circuit when $X_L = X_C$, or $2\pi fL = 1/2\pi fC$. Thus, the resonance frequency is given by

$$f_0 = \frac{1}{2\pi\sqrt{LC}}$$

To have $f_0 = f_{min} = 500 \text{ kHz} = 5.00 \times 10^5 \text{ Hz}$, the required capacitance in the tuning circuit is

$$C = C_{max} = \frac{1}{4\pi^2 L\left(f_{0,min}\right)^2} = \frac{1}{4\pi^2 \left(2.0 \times 10^{-6} \text{ H}\right)\left(5.00 \times 10^5 \text{ Hz}\right)^2}$$

or $C = C_{max} = 5.1 \times 10^{-8} \text{ F} = 51 \text{ nF}$ ◊

If the station is to be tuned to the maximum frequency in the AM radio band $\left(f_{max} = 1\,600 \text{ kHz} = 1.6 \times 10^6 \text{ Hz}\right)$, the capacitance in the tuning circuit for this radio must be

$$C = C_{min} = \frac{1}{4\pi^2 L\left(f_{0,max}\right)^2} = \frac{1}{4\pi^2 \left(2.0 \times 10^{-6} \text{ H}\right)\left(1.60 \times 10^6 \text{ Hz}\right)^2}$$

or $C = C_{min} = 4.9 \times 10^{-8} \text{ F} = 4.9 \text{ nF}$ ◊

---

45. An AC power generator produces 50 A (rms) at 3 600 V. The voltage is stepped up to 100 000 V by an ideal transformer, and the energy is transmitted through a long-distance power line that has a resistance of 100 Ω. What percentage of the power delivered by the generator is dissipated as heat in the power line?

Solution

The voltage applied to the primary coil of the transformer by the generator is $\Delta V_{1,\mathrm{rms}} = 3\ 600$ V, and the rms current in the primary coil is $I_{1,\mathrm{rms}} = 50$ A. The rms voltage across the secondary coil of the transformer is $\Delta V_{2,\mathrm{rms}} = 100\ 000$ V.

In an ideal transformer, **the power input to the primary equals the power output at the secondary,** or

$$I_{1,\mathrm{rms}}\Delta V_{1,\mathrm{rms}} = I_{2,\mathrm{rms}}\Delta V_{2,\mathrm{rms}}$$

Thus, the current in the circuit connected to the secondary of the transformer must be

$$I_{2,\mathrm{rms}} = \frac{I_{1,\mathrm{rms}}\Delta V_{1,\mathrm{rms}}}{\Delta V_{2,\mathrm{rms}}} = \frac{(50\text{ A})(3\ 600\text{ V})}{100\ 000\text{ V}} = 1.8\text{ A}$$

and the power loss in the resistance of the long-distance distribution line is

$$\mathcal{P}_{\mathrm{loss,line}} = I_{2,\mathrm{rms}}^2 R_{\mathrm{line}} = (1.8\text{ A})^2 (100\ \Omega) = 3.2 \times 10^2\text{ W}$$

The power delivered to the primary of the transformer by the generator is

$$\mathcal{P}_{\mathrm{input}} = I_{1,\mathrm{rms}}\Delta V_{1,\mathrm{rms}} = (50\text{ A})(3\ 600\text{ V}) = 1.8 \times 10^5\text{ W}$$

The percentage of the input power that is lost in the distribution line is therefore

$$\%\text{ Loss} = \frac{\mathcal{P}_{\mathrm{loss,line}}}{\mathcal{P}_{\mathrm{input}}} \times 100\% = \left(\frac{3.2 \times 10^2\text{ W}}{1.8 \times 10^5\text{ W}}\right) \times 100\% = 0.18\%$$

◊

53. Oxygenated hemoglobin absorbs weakly in the red (hence its red color) and strongly in the near infrared, while deoxygenated hemoglobin has the opposite absorption. This fact is used in a "pulse oximeter" to measure oxygen saturation in arterial blood. The device clips onto the end of a person's finger and has two light-emitting diodes [a red (660 nm) and an infrared (940 nm)] and a photocell that detects the amount of light transmitted through the finger at each wavelength. (a) Determine the frequency of each of these light sources. (b) If 67% of the energy of the red source is absorbed in the blood, by what factor does the amplitude of the electromagnetic wave change? *Hint:* The intensity of the wave is equal to the average power per unit area as given by Equation (21.28).

Solution

(a) The speed of electromagnetic waves in vacuum is $c = 3.00 \times 10^8$ m/s, so the relation between the wavelength, frequency, and speed of propagation of such waves is

$$\lambda f = c \quad \text{or} \quad f = c/\lambda$$

Red light with wavelength $\lambda_{red} = 660$ nm $= 660 \times 10^{-9}$ m $= 6.60 \times 10^{-7}$ m, is an electromagnetic wave having a frequency of

$$f_{red} = \frac{c}{\lambda_{red}} = \frac{3.00 \times 10^8 \text{ m}}{6.60 \times 10^{-7} \text{ m}} = 4.55 \times 10^{14} \text{ Hz} \qquad \Diamond$$

The infrared radiation is an electromagnetic wave with wavelength $\lambda_{IR} = 940$ nm $= 9.40 \times 10^{-7}$ m and frequency

$$f_{IR} = \frac{c}{\lambda_{IR}} = \frac{3.00 \times 10^8 \text{ m}}{9.4 \times 10^{-7} \text{ m}} = 3.19 \times 10^{14} \text{ Hz} \qquad \Diamond$$

(b) If 67% of the incident intensity, I_i, of the electromagnetic wave is absorbed in passing through the blood, the transmitted intensity is $I_t = I_i - 0.67 I_i = 0.33 I_i$.

The intensity (power per unit area) of an electromagnetic wave is proportional to the square of the amplitude of either the electric or magnetic component of the wave $[I = (1/2\mu_o c)E_{max}^2 = (c/2\mu_o c)B_{max}^2]$. Thus, if $I_t/I_i = 0.33$, we have

$$\frac{E_{max,t}}{E_{max,i}} = \frac{B_{max,t}}{B_{max,i}} = \sqrt{\frac{I_t}{I_i}} = \sqrt{0.33} = 0.57 \qquad \Diamond$$

That is, the transmitted amplitude of either the electric or magnetic component of the wave is 57% of the incident amplitude of that component. $\qquad \Diamond$

63. Infrared spectra are used by chemists to help identify an unknown substance. Atoms in a molecule that are bound together by a particular bond vibrate at a predictable frequency, and light at that frequency is absorbed strongly by the atom. In the case of the $C=O$ double bond, for example, the oxygen atom is bound to the carbon by a bond that has an effective spring constant of 2 800 N/m. If we assume that the carbon atom remains stationary (it is attached to other atoms in the molecule), determine the resonant frequency of this bond and the wavelength of light that matches that frequency. Verify that this wavelength lies in the infrared region of the spectrum. (The mass of an oxygen atom is 2.66×10^{-26} kg.)

Solution

Resonance will occur in a oscillating system when the system is excited by a periodic driving force having a frequency equal to the natural frequency of oscillation of the system. The natural frequency of oscillation of an object of mass m on the end of a spring having force constant k is given by

$$f_0 = \frac{1}{2\pi}\sqrt{\frac{k}{m}}$$

Thus, the resonant frequency for the $C=O$ double bond will be

$$f_0 = \frac{1}{2\pi}\sqrt{\frac{k_{\text{effective}}}{m_{\substack{\text{oxygen} \\ \text{atom}}}}} = \frac{1}{2\pi}\sqrt{\frac{2\ 800\ \text{N/m}}{2.66 \times 10^{-26}\ \text{kg}}} = 5.2 \times 10^{13}\ \text{Hz} \qquad \Diamond$$

The wavelength of light whose electric and magnetic fields oscillate at this frequency is

$$\lambda_0 = \frac{c}{f_0} = \frac{3.00 \times 10^8\ \text{m/s}}{5.2 \times 10^{13}\ \text{Hz}} = 5.8 \times 10^{-6}\ \text{m} = 5.8\ \mu\text{m} \qquad \Diamond$$

The infrared region of the electromagnetic spectrum includes wavelengths ranging from $\lambda_{\text{max}} \approx 1$ mm down to $\lambda_{\text{min}} = 700$ nm $= 0.7\ \mu$m. Thus, the radiation needed to produce resonance of the $C=O$ double bond lies in the infrared portion of the electromagnetic spectrum. $\qquad \Diamond$

71. As a way of determining the inductance of a coil used in a research project, a student first connects the coil to a 12.0-V battery and measures a current of 0.630 A. The student then connects the coil to a 24.0-V (rms), 60.0-Hz generator and measures an rms current of 0.570 A. What is the inductance?

Solution

When a DC voltage source (such as a battery) is connected to the coil, the inductive reactance of the coil is zero and only the resistance of its windings is effective in limiting the current through it. Thus, the resistance of this coil must be

$$R = \frac{\Delta V_{DC}}{I_{DC}} = \frac{12.0 \text{ V}}{0.630 \text{ A}} = 19.0 \text{ } \Omega$$

When an AC voltage is applied to the coil, both its inductive reactance and its resistance play a role in limiting the current. The impedance of the coil to the applied AC voltage is

$$Z = \frac{\Delta V_{rms}}{I_{rms}} = \frac{24.0 \text{ V}}{0.57 \text{ A}} = 42.1 \text{ } \Omega$$

Since, with zero capacitive reactance present, the impedance is given by $Z = \sqrt{R^2 + X_L^2}$, the inductive reactance of this coil at a frequency of $f = 60.0$ Hz is

$$X_L = \sqrt{Z^2 - R^2} = \sqrt{(42.1 \text{ } \Omega)^2 - (19.0 \text{ } \Omega)^2} = 37.6 \text{ } \Omega$$

and the self-inductance of the coil is

$$L = \frac{X_L}{2\pi f} = \frac{37.6 \text{ } \Omega}{2\pi (60.0 \text{ Hz})} = 9.97 \times 10^{-2} \text{ H} = 99.7 \text{ mH}$$ ◊

75. One possible means of achieving space flight is to place a perfectly reflecting aluminized sheet into Earth's orbit and to use the light from the Sun to push this solar sail. Suppose such a sail, of area 6.00×10^4 m² and mass 6 000 kg, is placed in orbit facing the Sun. (a) What force is exerted on the sail? (b) What is the sail's acceleration? (c) How long does it take for this sail to reach the Moon, 3.84×10^8 m away? Ignore all gravitational effects, and assume a solar intensity of 1 340 W/m². *Hint*: The radiation pressure by a reflected wave is given by 2(average power per unit area/c).

Solution

When electromagnetic radiation reflects (at normal incidence) from a perfectly reflecting surface, the momentum imparted to that surface in time Δt is $\Delta p = 2U/c$. Here, U is the energy delivered to the surface during Δt, and c is the speed of light.

From the impulse-momentum theorem, the average force exerted on the surface during this time is

$$F_{av} = \frac{\text{Impulse}}{\Delta t} = \frac{\Delta p}{\Delta t} = \frac{2U}{c(\Delta t)}$$

The intensity, I, of the incident radiation is the power transported per unit area, so the energy delivered to a surface of area A in time Δt is

$$U = \mathcal{P}_{av}(\Delta t) = (I \cdot A)(\Delta t)$$

and average force on the surface is $\qquad F_{av} = \frac{2(I \cdot A)(\Delta t)}{c(\Delta t)} = \frac{2IA}{c}$

(a) If $I = 1\,340$ W/m², the average force on a perfectly reflecting sail of area $A = 6.00 \times 10^4$ m² is

$$F_{av} = \frac{2(1\,340 \text{ W/m}^2)(6.00 \times 10^4 \text{ m}^2)}{3.00 \times 10^8 \text{ m/s}} = 0.536 \text{ N} \qquad \lozenge$$

(b) From Newton's second law, the average acceleration is

$$a_{av} = \frac{F_{av}}{m} = \frac{0.536 \text{ N}}{6\,000 \text{ kg}} = 8.93 \times 10^{-5} \text{ m/s}^2 \qquad \lozenge$$

(c) From the uniformly accelerated motion equation $\Delta x = v_0 t + \frac{1}{2}at^2$, with $v_0 = 0$, the time required to reach the moon is

$$t = \sqrt{\frac{2(\Delta x)}{a_{av}}} = \sqrt{\frac{2(3.84 \times 10^8 \text{ m})}{8.93 \times 10^{-5} \text{ m/s}^2}} = 2.93 \times 10^6 \text{ s}\left(\frac{1 \text{ day}}{86\,400 \text{ s}}\right) = 33.9 \text{ days} \qquad \lozenge$$

22

Reflection and Refraction of Light

NOTES FROM SELECTED CHAPTER SECTIONS

22.1 The Nature of Light

Thomas Young showed that light exhibits interference behavior, Maxwell predicted that light is a form of high-frequency electromagnetic waves, and Einstein explained the photoelectric effect on the assumption that light is composed of "corpuscles" or discontinuous quanta of energy called photons.

In view of these developments we conclude that:

Light must be regarded as having a dual nature; in some cases light acts as a wave, and in others it acts as a particle.

Light travels in a straight line in a homogeneous medium until it strikes a boundary between two different materials.

Light beams can be represented by a technique called the ray approximation in which a ray of light is shown as a line drawn along the direction of travel of the beam.

22.2 Reflection and Refraction

22.3 The Law of Refraction

Specular reflection occurs when light is reflected from a smooth surface; **diffuse reflection** occurs at rough surfaces. In this and following chapters reflection will be considered to be specular reflection.

When light is refracted:

The path of a light ray through a refracting surface is reversible.

The index of refraction is characteristic of a particular material.

The frequency does not change as light travels from one medium into another.

A line drawn perpendicular to a surface at the point where an incident ray strikes the surface is called the normal line. *Angles of incidence, reflection, and refraction are measured relative to the normal.* The incident, reflected, and refracted rays, and the normal line all lie in the same plane, and $\theta_1 = \theta_1'$.

For a given angle of incidence, the angle of refraction depends on the optical properties of the media above and below the boundary. The figure above illustrates a situation in which the speed of light is greater in the medium above the boundary than in the medium below the boundary. In such cases the angle of refraction, as shown, will be smaller than the angle of incidence.

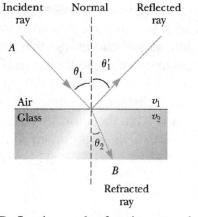

Reflection and refraction at a plane surface

θ_1 = angle of incidence

θ_1' = angle of reflection

θ_2 = angle of refraction

22.4 Dispersion and Prisms

An important property of the index of refraction (n) is that its value in a refractive material depends on the wavelength. This phenomenon is called dispersion. Since n is a function of the wavelength, when a light beam is incident on the surface of a refracting material, different wavelengths are bent at different angles. *The index of refraction of a given material decreases with increasing wavelength.* This means that blue light bends more than red light, when passing into a refracting material. In a prism, dispersion occurs at one surface as light enters the prism and again at a second surface as it leaves.

22.6 Huygens's Principle

A wave front is a surface which passes through those points in a wave that have both the same phase and amplitude. Huygens's technique for determining the successive positions of a wave front is based on the following geometric construction: All points on a wave front can be considered as point sources for the production of spherical secondary waves (wavelets). The wavelets propagate forward with speeds characteristic of waves in the particular medium. At any later time, the new position of the wave front is found by constructing a surface tangent to the set of wavelets.

22.7 Total Internal Reflection

When light is incident on a boundary between two media, some of the light is always reflected; the remaining light is refracted into the second medium. Total internal reflection is possible only when the light is initially traveling in the medium of greater index of refraction ($n_1 > n_2$). As the angle of incidence increases, the intensity of the reflected beam increases and that of the refracted beam decreases. If $n_1 > n_2$ there exists an angle of incidence called the critical angle, θ_c, beyond which no refraction occurs. When $\theta_1 \geq \theta_c$ the incident light is totally reflected back into the medium of greater index of refraction and the intensity of the refracted beam is zero.

EQUATIONS AND CONCEPTS

The law of reflection states that the angle of reflection (the angle measured between the reflected ray and the normal) equals the angle of incidence (the angle between the incident ray and the normal). *The incident ray, reflected ray, and normal line are in the same plane.*

$$\theta_1' = \theta_1 \tag{22.2}$$

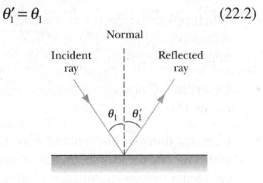

The **index of refraction** of a transparent medium equals the ratio of the speed of light in vacuum to the speed of light in the medium. The index of refraction of a given medium can be expressed as the ratio of the wavelength of light in vacuum to the wavelength in that medium. *The frequency of a wave is characteristic of the source; as light travels from one medium into another of different index of refraction, the frequency remains constant, but the wavelength changes.*

$$n \equiv \frac{\text{speed of light in vacuum}}{\text{speed of light in medium}} = \frac{c}{v} \tag{22.4}$$

$$n = \frac{\lambda_0}{\lambda_n} \tag{22.7}$$

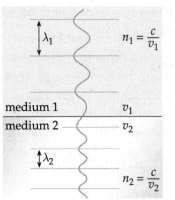

Snell's law of refraction can be expressed in terms of the speeds of light in the media on either side of the refracting surface or in terms of the indices of refraction of the two media. *As illustrated in the figure, the angles θ_1 and θ_2 are measured from the normal line to the respective ray.*

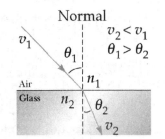

$$\frac{\sin\theta_2}{\sin\theta_1} = \frac{v_2}{v_1} = \text{constant} \tag{22.3}$$

$$n_1 \sin\theta_1 = n_2 \sin\theta_2 \tag{22.8}$$

The **critical angle** is the minimum angle of incidence for which total internal reflection can occur. *Total internal reflection is possible only when a light ray is directed from a medium of high index of refraction into a medium of lower index of refraction.*

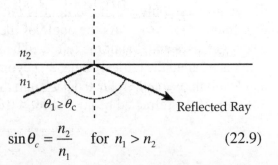

$$\sin\theta_c = \frac{n_2}{n_1} \quad \text{for } n_1 > n_2 \tag{22.9}$$

REVIEW CHECKLIST

- When refraction occurs at a plane surface, use Snell's law to calculate the index of refraction of either medium, the angle of refraction, or the angle of incidence when the other variables in the equation are known. (Section 22.3)

- Describe the process of dispersion of a beam of white light as it passes through a prism. (Sections 22.4 and 22.5)

- Describe the conditions under which total internal reflection of a light ray is possible, and calculate the critical angle for internal reflection at a boundary between two optical media of known indices of refraction. Describe the application of internal reflection to fiber optics techniques. (Section 22.7)

SOLUTIONS TO SELECTED END-OF-CHAPTER PROBLEMS

3. Find the energy of (a) a photon having a frequency of 5.00×10^{17} Hz, and (b) a photon having a wavelength of 3.00×10^2 nm. Express your answers in units of eV, noting that $1 \text{ eV} = 1.60 \times 10^{-19}$ J.

Solution

(a) The energy of a photon is given by

$$E = hf$$

where $h = 6.63 \times 10^{-34}$ J·s is known as Planck's constant, and f is the frequency of the electromagnetic wave associated with that photon. A photon having a frequency of 5.00×10^{17} Hz will have an energy, expressed in units of electron volts, of

$$E = hf = \left(6.63 \times 10^{-34} \text{ J·s}\right)\left(5.00 \times 10^{17} \text{ Hz}\right)\left(\frac{1 \text{ eV}}{1.60 \times 10^{-19} \text{ J}}\right) = 2.07 \times 10^3 \text{ eV} \quad \Diamond$$

(b) The relation between the speed, frequency, and wavelength in vacuum of an electromagnetic wave is $\lambda f = c$ or $f = c/\lambda$, where $c = 3.00 \times 10^8$ m/s is the speed of light in vacuum. Therefore, the energy of a photon can also be written as

$$E = \frac{hc}{\lambda}$$

The energy of a photon associated with an electromagnetic wave having a wavelength in vacuum of $\lambda = 3.00 \times 10^2$ nm is

$$E = \frac{hc}{\lambda} = \frac{\left(6.63 \times 10^{-34} \text{ J·s}\right)\left(3.00 \times 10^8 \text{ m/s}\right)}{3.00 \times 10^2 \text{ nm}}\left(\frac{1 \text{ nm}}{10^{-9} \text{ m}}\right) = 6.63 \times 10^{-19} \text{ J}$$

and, in units of electron volts,

$$E = \left(6.63 \times 10^{-19} \text{ J}\right)\left(\frac{1 \text{ eV}}{1.60 \times 10^{-19} \text{ J}}\right) = 4.14 \text{ eV} \quad \Diamond$$

11. A laser beam is incident at an angle of 30.0° to the vertical onto a solution of corn syrup in water. If the beam is refracted to 19.24° to the vertical, (a) what is the index of refraction of the syrup solution? Suppose the light is red, with wavelength 632.8 nm in a vacuum. Find its (b) wavelength, (c) frequency, and (d) speed in the solution.

Solution

(a) Applying Snell's law at the point where the beam crosses the boundary between the air and the syrup solution gives

$$n_{air} \sin 30.0° = n_{solution} \sin(19.24°)$$

$$\text{or} \quad n_{solution} = \frac{(1.00)\sin 30.0°}{\sin(19.24°)} = 1.52 \qquad \lozenge$$

(b) In a medium having refractive index n, light whose wavelength in a vacuum is λ_0 will have a wavelength of $\lambda_n = \lambda_0/n$. Thus,

$$\lambda_{solution} = \frac{\lambda_0}{n_{solution}} = \frac{632.8 \text{ nm}}{1.52} = 416 \text{ nm} \qquad \lozenge$$

(c) From $v = \lambda f$, we find

$$f = \frac{v_{solution}}{\lambda_{solution}} = \frac{c/n_{solution}}{\lambda_0/n_{solution}} = \frac{c}{\lambda_0} = \frac{3.00 \times 10^8 \text{ m/s}}{632.8 \times 10^{-9} \text{ m}} = 4.74 \times 10^{14} \text{ Hz} \qquad \lozenge$$

(d) The speed of light in the syrup solution is

$$v_{solution} = \frac{c}{n_{solution}} = \frac{3.00 \times 10^8 \text{ m/s}}{1.52} = 1.97 \times 10^8 \text{ m/s} \qquad \lozenge$$

---

17. How many times will the incident beam shown in Figure P22.17 be reflected by each of the parallel mirrors?

Solution

Each time the light beam reflects from a mirror, the angle of incidence and the angle of reflection are equal. The distance between successive points of reflection on either the right-side mirror or the left-side mirror is

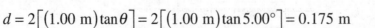

Modified Figure P22.17

$$d = 2\big[(1.00 \text{ m})\tan\theta\big] = 2\big[(1.00 \text{ m})\tan 5.00°\big] = 0.175 \text{ m}$$

Notice that the incident beam just misses the lower end of the right-side mirror, leaving the full length of this mirror available for reflections. Each reflection from this mirror "uses up" a length d of the mirror, and the total number of reflections which may be completed in the full 1.00-m length is

$$N_{\text{right}} = \frac{1.00 \text{ m}}{d} = \frac{1.00 \text{ m}}{0.175 \text{ m}} = 5.71 \quad (\text{or 5 full reflections}) \qquad \Diamond$$

The first reflection from the left-side mirror occurs a distance $d/2$ above the lower edge, leaving a mirror length of $(1.00 \text{ m} - d/2)$ to accommodate additional reflections. The number of *additional* reflections which may occur is

$$n_{\text{additional}} = \frac{1.00 \text{ m} - d/2}{d} = \frac{0.913 \text{ m}}{0.175 \text{ m}} = 5.22 \quad (\text{or 5 full reflections})$$

Thus, the total number of reflections from this mirror will be

$$N_{\text{left}} = 1 + n_{\text{additional}} = 1 + 5 = 6 \text{ reflections from the left side mirror} \qquad \Diamond$$

---

19. The light beam shown in Figure P22.19 makes an angle of 20.0° with the normal line NN' in the linseed oil. Determine the angles θ and θ'. (The refractive index for linseed oil is 1.48.)

Solution

First, we apply Snell's law to the refraction at the air-linseed oil boundary. To do this, realize that the upper and lower surfaces of the linseed oil layer are parallel to each other.

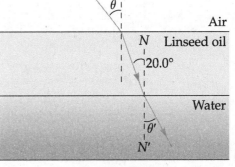

Figure P22.19

Thus, the two dashed normal lines shown in Figure P22.19 are parallel to each other, and the angle of refraction at the air-linseed oil boundary is 20.0°. Snell's law then gives $n_{\text{air}} \sin\theta = n_{\text{oil}} \sin 20.0°$, or

$$\theta = \sin^{-1}\left[\frac{n_{\text{oil}} \sin 20.0°}{n_{\text{air}}}\right] = \sin^{-1}\left[\frac{(1.48)\sin 20.0°}{1.00}\right] = 30.4° \qquad \Diamond$$

Application of Snell's law at the linseed oil-water boundary gives

$$n_{\text{water}} \sin\theta' = n_{\text{oil}} \sin 20.0°$$

or

$$\theta' = \sin^{-1}\left[\frac{n_{\text{oil}} \sin 20.0°}{n_{\text{water}}}\right] = \sin^{-1}\left[\frac{(1.48)\sin 20.0°}{1.333}\right] = 22.3° \qquad \Diamond$$

23. A person looking into an empty container is able to see the far edge of the container's bottom as in Figure P22.23a. The height of the container is h and its width is d. When the container is completely filled with a fluid of index of refraction n, the person can see a coin at the middle of the container's bottom as in Figure P22.23b.

(a) Show that the ratio h/d is given by

$$\frac{h}{d} = \sqrt{\frac{n^2 - 1}{4 - n^2}}$$

(b) If the container has a width of 8.0 cm and is filled with water, use the expression above to find the height of the container.

Solution

(a) Before the container is filled, the ray's path is as shown in Figure (a) at the right. From this figure, observe that

$$\sin\theta_1 = \frac{d}{s_1} = \frac{d}{\sqrt{h^2 + d^2}} = \frac{1}{\sqrt{(h/d)^2 + 1}}$$

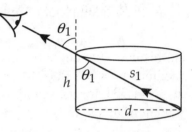

(a)

After the container is filled, the ray's path is shown in Figure (b). From this figure, we find that

$$\sin\theta_2 = \frac{d/2}{s_2} = \frac{d/2}{\sqrt{h^2 + (d/2)^2}} = \frac{1}{\sqrt{4(h/d)^2 + 1}}$$

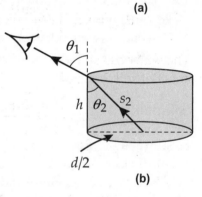

(b)

From Snell's law, $n_{air}\sin\theta_1 = n\sin\theta_2$, or

$$\frac{1.00}{\sqrt{(h/d)^2 + 1}} = \frac{n}{\sqrt{4(h/d)^2 + 1}}$$

and $4(h/d)^2 + 1 = n^2(h/d)^2 + n^2$. Simplifying, this gives $(4 - n^2)(h/d)^2 = n^2 - 1$

or $\quad \dfrac{h}{d} = \sqrt{\dfrac{n^2 - 1}{4 - n^2}}$ ◊

(b) If $d = 8.0$ cm and $n = n_{water} = 1.333$, then

$$h = (8.0\text{ cm})\sqrt{\frac{(1.333)^2 - 1}{4 - (1.333)^2}} = 4.7\text{ cm}$$ ◊

29. The index of refraction for red light in water is 1.331, and that for blue light is 1.340. If a ray of white light enters the water at an angle of incidence of 83.00°, what are the underwater angles of refraction for the blue and red components of the light?

Solution

As light travels from one medium having index of refraction n_1 into a second medium with index of refraction n_2, Snell's law gives the relation between the indices of refraction, the angle of incidence and the angle of refraction. This law states that

$$n_1 \sin\theta_1 = n_2 \sin\theta_2$$

and the angle of refraction is found to be

$$\theta_2 = \sin^{-1}\left(\frac{n_1 \sin\theta_1}{n_2}\right)$$

In this case, the angle of incidence is $\theta_1 = 83.00°$, and the index of refraction of the first medium is $n_1 = 1.000$, for both the red and blue light. With an index of refraction of $n_{2,red} = 1.331$ for the red light in water, the underwater angle of refraction for the red component of this light is

$$\theta_{2,red} = \sin^{-1}\left(\frac{n_1 \sin\theta_1}{n_{2,red}}\right) = \sin^{-1}\left(\frac{(1.000)\sin(83.00°)}{1.331}\right) = 48.22° \qquad \Diamond$$

The water has an index of refraction of $n_{2,blue} = 1.340$ for blue light, giving an underwater angle of refraction of

$$\theta_{2,blue} = \sin^{-1}\left(\frac{n_1 \sin\theta_1}{n_{2,blue}}\right) = \sin^{-1}\left(\frac{(1.000)\sin(83.00°)}{1.340}\right) = 47.79° \qquad \Diamond$$

---

33. A ray of light strikes the midpoint of one face of an equiangular (60°–60°–60°) glass prism ($n = 1.5$) at an angle of incidence of 30°. (a) Trace the path of the light ray through the glass, and find the angles of incidence and refraction at each surface. (b) If a small fraction of light is also reflected at each surface, find the angles of reflection at the surfaces.

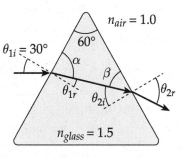

Solution

(a) The angle of incidence at the first surface is given to be $\theta_{1i} = 30°$. $\qquad \Diamond$

Snell's law then gives the angle of refraction at the first surface as:

$$\theta_{1r} = \sin^{-1}\left(\frac{n_{air} \sin\theta_{1i}}{n_{glass}}\right) = \sin^{-1}\left(\frac{1.0\sin 30°}{1.5}\right) = 19.47°$$

or rounding to 2 significant figures: $\theta_{1r} = 19°$ ◊

From the sketch, observe that $\theta_{1r} + \alpha = 90°$, or $\alpha = 90° - \theta_{1r}$

Also, $\alpha + \beta + 60° = 180°$ (sum of interior angles in a triangle). Therefore, the angle the refracted ray makes with the second surface is

$$\beta = 120° - \alpha = 120° - (90° - \theta_{1r}) = 30° + \theta_{1r} = 49.47°$$

The angle of incidence at the second surface is then

$$\theta_{2i} = 90° - \beta = 40.53°$$

or rounding to 2 significant figures: $\theta_{2i} = 41°$ ◊

Snell's law gives the angle of refraction at this surface as

$$\theta_{2r} = \sin^{-1}\left(\frac{n_{\text{glass}} \sin\theta_{2i}}{n_{\text{air}}}\right) = \sin^{-1}\left(\frac{1.5 \sin 40.53°}{1.0}\right) = 77°$$ ◊

(b)   From the law of reflection, the angle of reflection is equal to the angle of incidence at each of the surfaces. Thus,

$$\theta_{1,\,\text{reflection}} = \theta_{1i} = 30° \quad \text{and} \quad \theta_{2,\,\text{reflection}} = \theta_{2i} = 41°$$ ◊

---

40.   A beam of laser light with wavelength 612 nm is directed through a slab of glass having index of refraction 1.78. (a) For what minimum incident angle would a ray of light undergo total internal reflection? (b) If a layer of water is placed over the glass, what is the minimum angle of incidence on the glass-water interface that will result in total internal reflection at the water-air interface? (c) Does the thickness of the water layer or glass affect the result? (d) Does the index of refraction of the intervening layer affect the result?

Solution

(a)   The minimum angle of incidence for which total internal reflection occurs is the critical angle. At the critical angle, the angle of refraction is 90° as shown in the figure at the right. From Snell's law, $n_g \sin\theta_i = n_a \sin 90°$, the critical angle for the glass-air interface is found to be

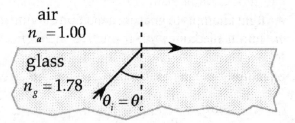

$$\theta_i = \theta_c = \sin^{-1}\left(\frac{n_a \sin 90°}{n_g}\right) = \sin^{-1}\left(\frac{1.00}{1.78}\right) = \boxed{34.2°}$$ ◊

(b) When the slab of glass has a layer of water on top, we want the angle of incidence at the water-air interface to equal the critical angle for that combination of media. At this angle, Snell's law gives

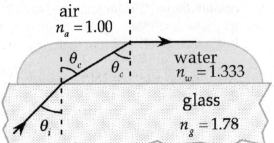

$$n_w \sin\theta_c = n_a \sin 90° = 1.00$$

and $\quad \sin\theta_c = 1.00/n_w$

Now, considering the refraction at the glass-water interface, Snell's law gives $n_g \sin\theta_i = n_w \sin\theta_c$. Combining this with the result for $\sin\theta_c$ from above, we find the required angle of incidence in the glass to be

$$\theta_i = \sin^{-1}\left(\frac{n_w \sin\theta_c}{n_g}\right) = \sin^{-1}\left(\frac{n_w\left(1.00/n_w\right)}{n_g}\right) = \sin^{-1}\left(\frac{1.00}{n_g}\right) = \sin^{-1}\left(\frac{1.00}{1.78}\right) = \boxed{34.2°} \quad \Diamond$$

(c) Notice that the thickness of the water layer was not needed in the calculations of the angles of incidence in part (b) above. Thus, that property of the water layer does not affect the result. While the thickness of the water layer has a role in determining the point where the light ray strikes the air-water interface, it does not affect the required angle of incidence. $\quad \Diamond$

(d) In the calculation of part (b), observe that the index of refraction of the water canceled. Thus, the refractive index of the intervening layer does not affect the required angle of incidence at the surface of the glass. This will always be true when the upper and lower surfaces of the intervening layer are parallel to each other. $\quad \Diamond$

---

43. The light beam in Figure P22.43 strikes surface 2 at the critical angle. Determine the angle of incidence, θ_i.

Solution

As light attempts to go from a medium having refractive index n_1 into a medium with refractive index n_2, where $n_1 > n_2$, total internal reflection occurs if the angle of incidence is equal to or greater than a critical angle, θ_c, given by

$$\sin\theta_c = \frac{n_2}{n_1}$$

Figure P22.43 (modified)

Where the light strikes surface 2 in Figure P22.43, the angle of incidence equals the critical angle and the indices of refraction are $n_1 = n_{glass}$, $n_2 = n_m$. Thus, we have that

$$\sin 42.0° = \frac{n_m}{n_{glass}}$$

At the upper surface of the prism, where the light is coming from the surrounding medium into the glass, Snell's law gives $n_m \sin\theta_i = n_{glass} \sin\theta_r$, or

$$\sin\theta_i = \frac{\sin\theta_r}{n_m/n_{glass}}$$

Combining this with the previous result gives

$$\sin\theta_i = \frac{\sin\theta_r}{\sin 42.0°}$$

Observe in Figure P22.43 that angle α is the complement of 42.0°, so $\alpha = 48.0°$. Also, $\alpha + \beta + 60.0° = 180°$ since they are the interior angles in a triangle. This gives

$$\beta = 180° - \alpha - 60.0° = 180° - 48.0° - 60.0° = 72.0°$$

Finally, observe that angle β is the complement of the angle of refraction at the upper surface, giving $\theta_r = 90.0° - \beta = 18.0°$. Thus, the angle of incidence at the upper surface of the prism must be

$$\theta_i = \sin^{-1}\left(\frac{\sin\theta_r}{\sin 42.0°}\right) = \sin^{-1}\left(\frac{\sin 18.0°}{\sin 42.0°}\right) = 27.5° \qquad \Diamond$$

---

49. As shown in Figure P22.49, a light ray is incident normal to one face of a 30°–60°–90° block of dense flint glass (a prism) that is immersed in water. (a) Determine the exit angle θ_4 of the ray. (b) A substance is dissolved in the water to increase the index of refraction. At what value of n_2 does total internal reflection cease at point P?

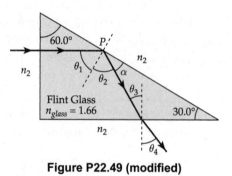

Figure P22.49 (modified)

Solution

(a) The horizontal ray strikes the first surface of the prism at normal incidence and passes with zero deviation into the glass. Notice that, inside the glass, this horizontal ray is parallel to the base of the prism, and hence makes a 30.0° angle with the diagonal face of the prism. This means that the angle of incidence at point P is $\theta_1 = 90.0° - 30.0° = 60.0°$. From the law of reflection, we see that $\theta_2 = \theta_1 = 60.0°$ and $\alpha = 90.0° - \theta_2 = 30.0°$. Then, since the sum of the interior angles of a triangle is 180°, we have $\alpha + (\theta_3 + 90.0°) + 30.0° = 180°$, or the angle of incidence at the lower surface of the prism is $\theta_3 = 180° - 120° - \alpha = 30.0°$.

With water as the surrounding medium ($n_2 = 1.333$), and $n_{glass} = 1.66$ for flint glass, Snell's law gives

$$\theta_4 = \sin^{-1}\left(\frac{n_{glass}\sin\theta_3}{n_2}\right) = \sin^{-1}\left(\frac{1.66\sin 30.0°}{1.333}\right) = 38.5° \qquad \Diamond$$

(b) Total internal reflection ceases at point P when the angle of incidence at this point equals the critical angle. That is, when

$$\sin\theta_1 = \sin\theta_c = \frac{n_2}{n_{\text{glass}}}$$

or $\quad n_2 = n_{\text{glass}}\sin\theta_1 = (1.66)\sin 60.0° = 1.44$ ◊

---

53. A piece of wire is bent through an angle θ. The bent wire is partially submerged in benzene (index of refraction = 1.50), so that, to a person looking along the dry part, the wire appears to be straight and makes an angle of 30.0° with the horizontal. Determine the value of θ.

Solution

The sketch at the right shows a light ray that leaves the lower end of the wire and travels parallel to the wire in the benzene. If the wire is to appear to be straight to a person looking along the dry part of the wire, the lower end of the wire must appear to be aligned with the dry portion of the wire, or lie along the dashed line in the sketch. Thus, when the shown ray from the lower end reaches the air-benzene interface, it must refract and travel parallel to the dry portion of the wire to enter the observer's eye.

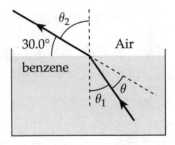

If the dry part of the wire, and hence the refracted ray in the sketch, are directed at 30.0° above the benzene surface, the angle of refraction is $\theta_2 = 90.0° - 30.0° = 60.0°$. Snell's law then gives the angle of incidence at the air-benzene interface as

$$\theta_1 = \sin^{-1}\left[\frac{n_{\text{air}}\sin\theta_2}{n_{\text{benzene}}}\right] = \sin^{-1}\left[\frac{(1.00)\sin 60.0°}{1.50}\right] = 35.3°$$

Since $\theta_2 = \theta_1 + \theta$ (vertical angles), the angle through which the wire is bent (or deviates from the dashed straight line path) is seen to be

$$\theta = \theta_2 - \theta_1 = 60.0° - 35.3° = 24.7°$$ ◊

59. Figure P22.59 shows the path of a beam of light through several layers with different indices of refraction. (a) If $\theta_1 = 30.0°$, what is the angle θ_2 of the emerging beam? (b) What must the incident angle θ_1 be in order to have total internal reflection at the surface between the medium with $n = 1.20$ and the medium with $n = 1.00$?

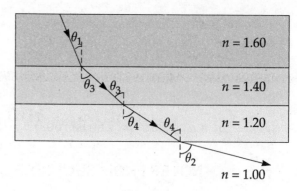

Figure P22.59 (modified)

Solution

First, we write Snell's law for the refraction at each of the boundaries.

At the boundary between $n = 1.60$ and $n = 1.40$: $1.60\sin\theta_1 = 1.40\sin\theta_3$ (1)

At the boundary between $n = 1.40$ and $n = 1.20$: $1.40\sin\theta_3 = 1.20\sin\theta_4$ (2)

At the boundary between $n = 1.20$ and $n = 1.00$: $1.20\sin\theta_4 = 1.00\sin\theta_2$ (3)

Since the right side of Equation (1) equals the left side of Equation (2), and the right side of Equation (2) equals the left side of Equation (3), we see that

$$1.60\sin\theta_1 = 1.00\sin\theta_2 \qquad (4)$$

Note that Equation (4) is exactly the same as we could have obtained from Snell's law by simply ignoring the presence of the second and third layers of this "sandwich." This will always be true if all the surfaces of the layers in the sandwich of materials are parallel to each other.

(a) If $\theta_1 = 30.0°$, Equation (4) gives $\theta_2 = \sin^{-1}\left(\dfrac{1.60\sin 30.0°}{1.00}\right) = 53.1°$ ◊

(b) At the critical angle of incidence at the $n = 1.20$ and $n = 1.00$ boundary, the refracted ray would be parallel to this boundary (that is, $\theta_2 = 90.0°$). From Equation (4), we find the minimum angle of incidence, $\left(\theta_1\right)_{\text{min}}$, which would yield total internal reflection at the lower boundary to be

$$\left(\theta_1\right)_{\text{min}} = \sin^{-1}\left(\frac{1.00\sin\theta_{2,\,\text{max}}}{1.60}\right) = \sin^{-1}\left(\frac{1.00\sin 90.0°}{1.60}\right) = 38.7° \qquad ◊$$

23

Mirrors and Lenses

NOTES FROM SELECTED CHAPTER SECTIONS

In equations and diagrams, the **object distance,** p, is the distance from the object to the mirror (or lens). The **image distance,** q, is the distance from the mirror (or lens) to the location of the image.

23.1 Flat Mirrors

Images formed by flat mirrors have the following properties:

- The image is as far behind the mirror as the object is in front.

- The image is unmagnified, virtual, and upright.

- The image has *apparent* left-right reversal.

23.2 Images Formed by Spherical Mirrors

A **spherical mirror** is a reflecting surface which has the shape of a segment of a sphere. If the inner surface is reflecting, the mirror is **concave;** if the outer surface of the sphere is reflecting, then the mirror is **convex.**

The **principal axis** of a spherical mirror is along a line drawn from the center (or apex) of the mirror to the **center of curvature** of the mirror. The **focal point** of a concave mirror is on the front side of the mirror at the point along the principal axis at which incoming parallel light rays converge or focus after reflection. For a convex mirror the focal point is on the back side of the mirror at the location from which incoming parallel rays appear to diverge after reflection. The **focal length** (the distance from the mirror to the focal point) is positive for a concave mirror and negative for a convex mirror.

A real image is formed at a point when reflected light rays actually pass through the point.

A virtual image is formed at a point when reflected light rays appear to diverge from the point.

23.3 Convex Mirrors and Sign Conventions

A convex mirror is a diverging mirror; light rays from an object (incident on the front side of the mirror) are reflected so as to appear to be coming from the image position (on the back side of the mirror). *Images of real objects formed by convex mirrors are always virtual, upright, and smaller than the object. This is true for all object distances.*

A concave mirror is a converging mirror. The nature of an image (real or virtual, upright or inverted, smaller or larger than the object) formed by a concave mirror depends on the object distance, p, relative to the focal length, f. Possible object locations and image outcomes are given in the table below:

If the object distance is...	Then the image will be
$p > 2f$	real, inverted, smaller than object
$p = 2f$	real, inverted, same size as object
$f < p < 2f$	real, inverted, larger than object
$p = f$	reflected rays are parallel, no image
$p < f$	virtual, upright, larger than object

A **ray diagram** is a graphical technique (scale drawing) that can be used to determine the location, nature, and relative size of an image formed by a spherical mirror. The positions of the object and focal point are shown along the principal axis of the mirror. **See the example ray diagrams for mirrors in Suggestions, Skills, and Strategies.**

The point of intersection of any two of the following three rays in a ray diagram for mirrors locates the image:

1. Ray #1 is drawn from the top of the object parallel to the principal axis and is reflected back through the focal point of a concave mirror. In the case of a convex mirror, the reflected ray appears to diverge from the focal point.

2. Ray #2 is drawn from the top of the object through the focal point of a concave mirror (toward the focal point of a convex mirror) and is reflected parallel to the principal axis.

3. Ray #3 is drawn from the top of the object through the center of curvature and is reflected back on itself.

The point at which the rays intersect in the case of a concave mirror (or the point from which they appear to diverge in the case of a convex mirror) locates the top of the image.

23.4 Images Formed by Refraction

A **real object** is one which is located on the front side of a refracting surface. A **virtual object** exists when the incident light rays are converging toward a point located on the back side of a refracting surface. The front side is defined as the side of the surface in which light rays originate. The location and character of an image formed by a refracting surface depends on the shape of the surface (convex, concave, or flat) and the relative values of the indexes of refraction of the media on the two sides of the refracting surface.

The following cases for real objects are illustrated in Suggestions, Skills, and Strategies:

(a)   An object located beyond the focal point on the front side of a convex surface where $n_1 < n_2$ (n_1 in the region of the incident ray and n_2 in the region of the refracted ray) will result in a real image on the back side of the refracting surface.

(b)   An object located on the front side of a concave surface where $n_1 < n_2$ will result in a virtual image on the front side of the refracting surface.

(c)   An object located on the front side of a convex surface where $n_1 > n_2$ will result in a virtual image on the front side of the refracting surface.

(d)   An object located on the front side of a flat surface will result in a virtual image on the front side of the surface, with $|q| < p$ if $n_2 < n_1$ and with $|q| > p$ if $n_2 > n_1$.

23.6 Thin Lenses

A converging lens is thicker in the middle than at the rim and has a positive focal length (when surrounded by a medium with an index of refraction smaller than that of the lens). A diverging lens is thicker at the rim and has a negative focal length.

A thin lens has two focal points as illustrated below. Note the reversal of locations of F_1 and F_2 for the two lens types.

The following three rays form the ray diagram for a thin lens:

1.   Ray #1 is drawn from the object parallel to the principal axis. After being refracted by the lens, this ray passes through focal point F_1 in the case of a converging lens and appears to diverge from F_1 in the case of a diverging lens.

2.   Ray #2 is drawn through the center of the lens. This ray continues in a straight line.

3.   Ray #3 is drawn through focal point F_2 (toward F_2 in the case of a diverging lens), and emerges from the lens parallel to the principal axis.

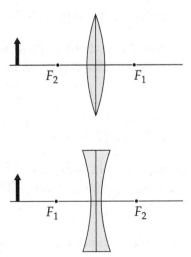

In the figure above, consider the arrow to be an object and construct a ray diagram for each of the two cases shown.

See the examples of ray diagrams for thin lenses in Suggestions, Skills, and Strategies.

When a combination of two lenses is used to form an image, the image of the first lens is treated as the object for the second lens. The overall magnification of the combination is the product of the magnifications of the separate lenses.

23.7 Lens and Mirror Aberrations

Aberrations are responsible for the formation of imperfect images by lenses and mirrors. **Spherical aberration** is due to the fact that parallel incident light rays that are different distances from the optical axis are focused at different points following refraction or reflection. **Chromatic aberration** arises from the fact that light of different wavelengths focuses at different points when refracted by a lens. This occurs because the index of refraction is a function of wavelength.

EQUATIONS AND CONCEPTS

Lateral magnification of a flat mirror is defined as the ratio of image height to the object height. *This ratio always has a value of +1 for a flat mirror since the upright image is always the same size as the object.*

$$M \equiv \frac{\text{image height}}{\text{object height}} = \frac{h'}{h} \qquad (23.1)$$

Lateral magnification of a spherical mirror can be expressed either as the ratio of the image size to the object size or as the negative of the ratio of the image distance to the object distance. *When the image is inverted both M and h' will have negative values.*

$$M = \frac{h'}{h} = -\frac{q}{p} \qquad (23.2)$$

The focal length of a spherical mirror is the distance from the vertex of the mirror to the focal point, *F*, located midway between the center of curvature and the vertex of the mirror. *For a concave mirror, the focal point is in front of the mirror, and f is positive. For a convex mirror, the focal point is back of the mirror, and f is negative.*

$$f = \frac{R}{2} \qquad (23.5)$$

R = radius of curvature

The mirror equation is used to determine the location of an image formed by reflection of paraxial rays from a spherical surface.

$$\frac{1}{p} + \frac{1}{q} = \frac{1}{f} \quad \begin{pmatrix} p = \text{object distance} \\ q = \text{image distance} \end{pmatrix} \qquad (23.6)$$

An image formed by a spherical refracting surface of radius R, separating two media with indices of refraction n_1 and n_2, can be real or virtual depending on the value of R. In Equation (23.7), n_1 is the index of refraction on the side of the incident rays, and n_2 is the index of refraction on the side of the refracted rays.

$$\frac{n_1}{p} + \frac{n_2}{q} = \frac{n_2 - n_1}{R} \tag{23.7}$$

$$M = \frac{h'}{h} = -\frac{n_1 q}{n_2 p} \tag{23.8}$$

A **flat refracting surface** $(R = \infty)$ will form an image on the same side of the surface as the object regardless of the relative values of n_1 and n_2. *You should review the example ray diagrams and sign conventions for refracting surfaces in* **Suggestions, Skills, and Strategies.**

$$q = -\frac{n_2}{n_1} p \tag{23.9}$$

Several **combinations of lens parameters and image characteristics for a thin lens** are shown in the following equations.

Features of the geometry of a thin lens are shown in the figures at right. A lens can be considered "thin" when the lens thickness is much less that the radii of the surfaces (R_1 and R_2). A given radius is positive if the center of curvature is back of the lens and negative if the center of curvature is in front of the lens. In the figure, R_1 is positive and R_2 is negative. A convex lens (bottom left) converges incident parallel light rays to the focal point (F_2). Parallel light rays entering a concave lens (bottom right) appear to diverge from the focal point (F_1).

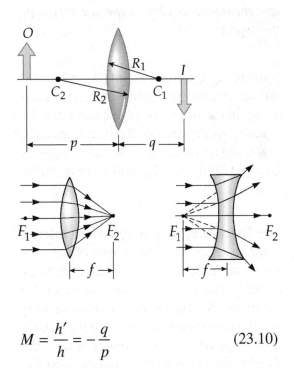

The **lateral magnification of a thin lens** as shown in Equation (23.10) has the same form as that of a spherical mirror.

$$M = \frac{h'}{h} = -\frac{q}{p} \tag{23.10}$$

The **thin lens equation** can be used to find the image location when the focal length and object distanced are known. The over-all magnification of an image formed by two optical elements (e.g., lens, mirror, or lens-mirror combination) equals the product of the magnifications due to the two individual elements.

$$\frac{1}{p}+\frac{1}{q}=\frac{1}{f} \qquad (23.11)$$

$$M = M_1 M_2$$

The **lens maker's equation** can be used to calculate the focal length of a thin lens in terms of the physical characteristics of the lens. *If the lens is surrounded by a medium other than air, the index of refraction given in Equation (23.12) must be the ratio of the index of refraction of the lens to that of the surrounding medium.*

$$\frac{1}{f}=(n-1)\left(\frac{1}{R_1}-\frac{1}{R_2}\right) \qquad (23.12)$$

For a convex lens R_1 is positive and R_2 is negative.

SUGGESTIONS, SKILLS, AND STRATEGIES

A major portion of this chapter is devoted to the development and presentation of equations which can be used to determine the location and nature of images formed by various optical components acting either singly or in combination. It is essential that these equations be used with the correct algebraic sign associated with each quantity involved. You must understand clearly the sign conventions for mirrors, refracting surfaces, and lenses. Sign conventions to be used with spherical mirrors, thin lenses, and refracting surfaces are summarized on the following pages.

SIGN CONVENTIONS FOR SPHERICAL MIRRORS

Equations: $\dfrac{1}{p} + \dfrac{1}{q} = \dfrac{1}{f} = \dfrac{2}{R}$ $\qquad M = \dfrac{h'}{h} = -\dfrac{q}{p}$

The front side of the mirror is the region on which light rays are incident and reflected.

p is + if the object is in front of the mirror (real object).
p is − if the object is in back of the mirror (virtual object).

q is + if the image is in front of the mirror (real image).
q is − if the image is in back of the mirror (virtual image).

Both f and R are + if the center of curvature is in front (concave mirror).
Both f and R are − if the center of curvature is in back (convex mirror).

If M is positive, the image is upright.
If M is negative, the image is inverted.

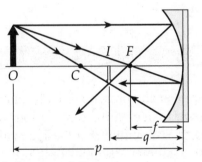

(a) Concave Mirror
$p > 2f$: $q+, f+, R+$
Image real, inverted, diminished

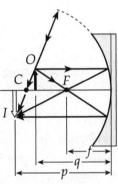

(b) Concave Mirror
$2f > p > f$: $q+, f+, R+$
Image real, inverted, enlarged

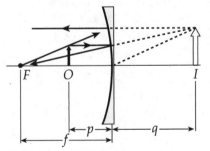

(c) Concave Mirror
$p < f$: $q-, f+, R+$
Image virtual, upright, enlarged

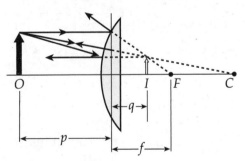

(d) Convex Mirror
$p+, q-, f-, R-$
Image virtual, upright, diminished
for any object distance

SIGN CONVENTIONS FOR REFRACTING SURFACES

Equations:

$$\frac{n_1}{p} + \frac{n_2}{q} = \frac{n_2 - n_1}{R} \qquad M = \frac{h'}{h} = -\frac{n_1 q}{n_2 p}$$

In the following table, the **front** side of the surface is the side **from which the light is incident.**

---

p is + if the object is in front of the surface (real object).
p is − if the object is in back of the surface (virtual object).

q is + if the image is in back of the surface (real image).
q is − if the image is in front of the surface (virtual image).

R is + if the center of curvature is in back of the surface.
R is − if the center of curvature is in front of the surface.

n_1 refers to the index of refraction of the first medium (before refraction).
n_2 is the index of refraction of the second medium (after refraction).

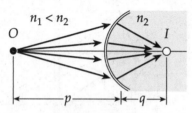

(a) p + (real object)
q + (real image)
R + (convex to incident light)

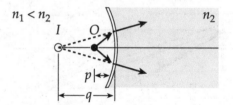

(b) p + (real object)
q − (virtual image)
R − (concave to incident light)

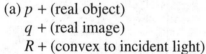
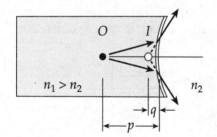

(c) p + (real object)
q − (virtual image)
R + (concave to incident light)

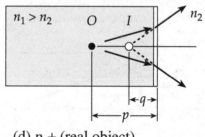

(d) p + (real object)
q − (virtual image)

SIGN CONVENTIONS FOR THIN LENSES

Equations:
$$\frac{1}{p} + \frac{1}{q} = \frac{1}{f} = (n-1)\left(\frac{1}{R_1} - \frac{1}{R_2}\right) \qquad M = \frac{h'}{h} = -\frac{q}{p}$$

In the following table, the **front** of the lens is the **side from which the light is incident.**

> p is + if the object is in front of the lens.
> p is − if the object is in back of the lens.
>
> q is + if the image is in back of the lens.
> q is − if the image is in front of the lens.
>
> f is + if the lens is thickest at the center.
> f is − if the lens is thickest at the edges.
>
> R_1 and R_2 are + if the center of curvature is in back of the lens.
> R_1 and R_2 are − if the center of curvature is in front of the lens.

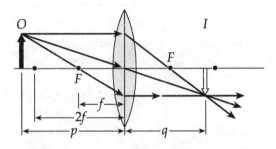

(a) Converging Lens ($f > 0$)
   $p > 2f$: $p+$, $q+$
   Image is real, inverted, diminished

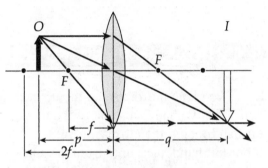

(b) Converging Lens ($f > 0$)
   $2f > p > f$: $p+$, $q+$
   Image is real, inverted, enlarged

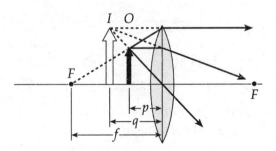

(c) Converging Lens ($f > 0$)
   $p < f$: $p+$, $q-$
   Image is virtual, upright, enlarged

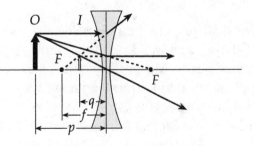

(d) Diverging Lens ($f < 0$)
   $p+$, $q-$
   Image is virtual, upright, diminished

REVIEW CHECKLIST

- You should know the sign conventions used with mirrors, lenses, and refracting surfaces. (Sections 23.3, 23.4, and 23.6)

- Understand the manner in which the algebraic signs associated with calculated quantities correspond to the nature of the image: real or virtual, upright or inverted. (Sections 23.3, 23.4, and 23.6)

- Calculate the location of the image of a specified object as formed by a flat mirror, spherical mirror, plane refracting surface, spherical refracting surface, or thin lens. Determine the magnification and character of the image in each case. (Sections 23.3, 23.4, and 23.6).

- Construct ray diagrams to determine the location and nature of the image of a given object when the geometric characteristics of the optical device (lens, refracting surface, or mirror) are known. (Sections 23.3, 23.4, and 23.6)

SOLUTIONS TO SELECTED END-OF-CHAPTER PROBLEMS

2.   Two plane mirrors stand facing each other, 3.0 m apart and a woman stands between them. The woman faces one of the mirrors from a distance of 1.0 m, with the palm of her left hand facing the closer mirror. (a) What is the apparent position of the closest image of her left hand, measured from the surface of the mirror in front of her? Does it show her palm or the back of her hand? (b) What is the position of the next image? Does it show her palm or the back of her hand? (c) Repeat for third image. (d) Which of the images are real and which are virtual?

Solution

We shall refer to the mirror located 1.0 m in front of the woman as M_F and the mirror located 2.0 m behind her as M_B. For any object located in front of a plane (or flat) mirror, the mirror forms an upright, virtual image that is the same size as the object, and is located as far behind the mirror as the object is in front of the mirror.

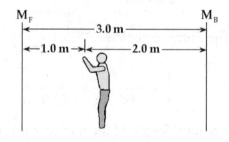

(a)   The closest image the woman sees is an image of her palm, located 1.0 m in front of the mirror M_F. This mirror forms an upright, virtual image $I_{palm,1}$ of her palm 1.0 m behind M_F.   ◊

(b) The back of the woman's hand is located 2.0 m in front of mirror M_B, which forms an image $I_{back,1}$ 2.0 m behind the mirror. Image $I_{back,1}$ is an upright, virtual image of the back of her hand, and it serves as an object for mirror M_F. This object is located $d = 2.0\ m + 3.0\ m = 5.0\ m$ in front of M_F, and this mirror forms an upright, virtual image, $I_{back,2}$, of the back of the hand, 5.0 m behind M_F. This is the second closest image seen by the woman.   ◊

(c) The image $I_{palm,1}$ [see part (a)] serves as an object for mirror M_B. Since this object is located distance $d' = 1.0\ m + 3.0\ m = 4.0\ m$ in front of M_B, the mirror forms an upright, virtual image $I_{palm,2}$ located 4.0 m behind M_B. Image $I_{palm,2}$ is located distance $d'' = 4.0\ m + 3.0\ m = 7.0\ m$ in front of mirror M_F, and serves as an object for that mirror. Mirror M_F then forms an upright, virtual image of the woman's palm at a distance of 7.0 m behind the mirror. This is the third closest image the woman can see.   ◊

(d) All images formed by either of these mirrors are virtual images.   ◊

---

11. A 2.00-cm-high object is placed 3.00 cm in front of a concave mirror. If the image is 5.00 cm high and virtual, what is the focal length of the mirror?

Solution

Virtual images formed by concave mirrors are upright images. Hence, the object height, h, and the image height, h', have the same sign, and the magnification is positive.

If $h' = 5.00$ cm when $h = 2.00$ cm, we have $M = \dfrac{h'}{h} = -\dfrac{q}{p}$

and $\qquad \dfrac{5.00\ \text{cm}}{2.00\ \text{cm}} = -\dfrac{q}{p}$ $\qquad$ or $\qquad q = -2.50\ p = -2.50(3.00\ \text{cm}) = -7.50\ \text{cm}$

The mirror equation $\qquad \dfrac{1}{p} + \dfrac{1}{q} = \dfrac{1}{f}$ $\qquad$ then gives

$$\dfrac{1}{f} = \dfrac{1}{3.00\ \text{cm}} + \dfrac{1}{-7.50\ \text{cm}} \qquad \text{or} \qquad \dfrac{1}{f} = \dfrac{-7.50\ \text{cm} + 3.00\ \text{cm}}{(3.00\ \text{cm})(-7.50\ \text{cm})}$$

The focal length of the mirror is given by

$$f = \dfrac{(3.00\ \text{cm})(-7.50\ \text{cm})}{-7.50\ \text{cm} + 3.00\ \text{cm}} = +5.00\ \text{cm} \qquad\qquad ◊$$

---

15. A man standing 1.52 m in front of a shaving mirror produces an inverted image 18.0 cm in front of it. How close to the mirror should he stand if he wants to form an upright image of his chin that is twice the chin's actual size?

Solution

When the man stands 1.52 m in front of the mirror $(p = +1.52$ m $= 152$ cm$)$, a real image is formed 18.0 cm in front of the mirror $(q = +18.0$ cm$)$. From the mirror equation, $1/f = 1/p + 1/q$, the focal length is found to be

$$f = \frac{pq}{p+q} = \frac{(152 \text{ cm})(18.0 \text{ cm})}{152 \text{ cm} + 18.0 \text{ cm}} = 16.1 \text{ cm}$$

If the mirror is now to form an upright image, the magnification will be positive. If this upright image is to be twice the size of the object, the magnitude of the magnification will be 2. Thus, we know that $M = +2$.

From this, we find $\quad M = -\dfrac{q}{p} = +2 \quad$ or $\quad q = -2p$

and the mirror equation gives $\quad \dfrac{1}{p} - \dfrac{1}{2p} = \dfrac{1}{f} \quad$ or $\quad \dfrac{1}{2p} = \dfrac{1}{f}$

The required object distance is then $\quad p = \dfrac{f}{2} = \dfrac{16.1 \text{ cm}}{2} = 8.05 \text{ cm}$ ◊

---

19. A spherical mirror is to be used to form an image, five times as tall as an object, on a screen positioned 5.0 m from the mirror. (a) Describe the type of mirror required. (b) Where should the mirror be positioned relative to the object?

Solution

(a) Since the image is formed on a screen located in front of the mirror, the image is real and the image distance is positive $(q = +5.0$ m$)$.

We could also have determined that the image is real by recognizing that the light reflecting from the mirror must actually come to a focus on the screen in order to form an image on it. Since real images are located where the light actually comes to a focus, while virtual images are located where the light the light has not traveled but still appears to diverge away from, we again conclude that the image is real.

Of the different types of mirrors (flat mirrors, concave mirrors, and convex mirrors), only a concave mirror will form a real image of a real object. Recognizing this, we conclude that the mirror used to form this image is a concave mirror. ◊

(b) With both $p > 0$ and $q > 0$, the magnification given by $M = -q/p$ will be negative, indicating that the image is inverted.

If the image is five times the size of the object, then

$$|M| = \frac{q}{p} = 5 \quad \text{and} \quad q = 5p$$

The object distance is then $p = q/5 = (5.0 \text{ m})/5 = 1.0 \text{ m}$, or the mirror should be 1.0 m from the object. ◊

If desired, we may now use the mirror equation $1/p + 1/q = 2/R$ to determine the radius of curvature of the mirror. This gives

$$R = \frac{2pq}{p+q} = \frac{2(1.0 \text{ m})(5.0 \text{ m})}{1.0 \text{ m} + 5.0 \text{ m}} = +1.7 \text{ m}$$

---

25. A transparent sphere of unknown composition is observed to form an image of the Sun on its surface opposite the Sun. What is the refractive index of the sphere material?

Solution

The image of the Sun is formed by the refraction of light at the spherical surface facing the Sun. The center of curvature of this surface and the image formed are both located on the back side of this surface. Thus, by the sign convention for refracting surfaces (see Table 23.2 in the textbook), both the radius of curvature and the image distance are positive. Since the image is formed on the side of the sphere opposite the Sun, the image distance is

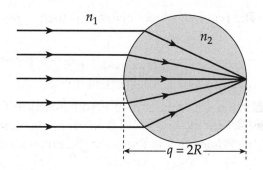

$$q = +(diameter \ of \ sphere) = +2R$$

When light refracts at a spherical surface, going from a medium with refractive index n_1 into a medium having refractive index n_2, the object distance, image distance, and radius of curvature are related by

$$\frac{n_1}{p} + \frac{n_2}{q} = \frac{n_2 - n_1}{R}$$

The sphere is surrounded by air, so $n_1 = 1.00$. Because the object distance (the Earth-Sun distance) is very large, $p \gg 1$ and $n_1/p \to 0$. Thus, we have

$$0 + \frac{n_2}{q} = \frac{n_2 - 1.00}{R} \quad \text{or, with} \quad q = 2R, \quad \frac{n_2}{2R} = \frac{n_2 - 1.00}{R}$$

This gives $\quad n_2 = 2n_2 - 2.00 \quad$ and $\quad n_2 = 2.00$ $\qquad\qquad\qquad$ ◊

Note that the radius of the sphere canceled, meaning that required index of refraction of the glass is the same regardless of the size of the sphere.

---

29. A contact lens is made of plastic with an index of refraction of 1.50. The lens has an outer radius of curvature of +2.00 cm and an inner radius of curvature of +2.50 cm. What is the focal length of the lens?

Solution

A contact lens is made to fit over the front of the eye and has a convex-concave shape as shown in the sketch at the right. As light travels from left to right through this lens, notice that the center of curvature of each of the refracting surfaces are located behind the surface. That is, the center of curvature C is located on the side of the surface the light is going toward as it crosses the surface, not on the side the light is coming from. Thus, by the sign convention of Table 23.2 in the textbook, the radius of curvature is positive for both surfaces of this lens.

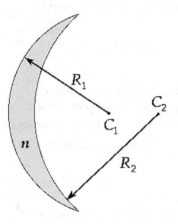

For the lens under consideration, $R_1 = +2.00$ cm, $R_2 = +2.50$ cm, and the index of the lens material $n = 1.50$. The lens maker's equation gives the focal length of a thin lens, surrounded by air, in terms of the radii of its surfaces and the index of refraction of the lens material as

$$\frac{1}{f} = (n-1)\left(\frac{1}{R_1} - \frac{1}{R_2}\right)$$

Thus, for the described lens

$$\frac{1}{f} = (1.50 - 1)\left(\frac{1}{2.00 \text{ cm}} - \frac{1}{2.50 \text{ cm}}\right) = (0.50)\left[\frac{2.50 \text{ cm} - 2.00 \text{ cm}}{(2.00 \text{ cm})(2.50 \text{ cm})}\right] = \frac{(0.50)^2}{5.00 \text{ cm}}$$

and the focal length is found to be

$$f = \frac{5.00 \text{ cm}}{0.25} = +20 \text{ cm} \qquad\qquad ◊$$

---

33. A diverging lens has a focal length of 20.0 cm. Locate the images for object distances of (a) 40.0 cm, (b) 20.0 cm, and (c) 10.0 cm. For each case, state whether the image is real or virtual and upright or inverted, and find the magnification.

Solution

The focal length of a diverging lens is negative, so the focal length of the given lens is $f = -20.0$ cm. For an object distance, p, the thin lens equation $1/p + 1/q = 1/f$ gives the image distance as

$$q = \frac{pf}{p-f}$$

and the magnification is

$$M = -\frac{q}{p}$$

The image is real if $q > 0$, and virtual if $q < 0$. Also, the image is upright if $M > 0$, and inverted if $M < 0$.

(a) When $p = +40.0$ cm, the image distance is

$$q = (40.0 \text{ cm})(-20.0 \text{ cm})/[40.0 \text{ cm} - (-20.0 \text{ cm})] = -13.3 \text{ cm}$$

and the magnification is $M = -(-13.3 \text{ cm})/(40.0 \text{ cm}) = +1/3$.

Thus, the image is virtual, located 13.3 cm in front of the lens, upright, and one-third the size of the object.   ◊

(b) When $p = +20.0$ cm, the image distance is

$$q = (20.0 \text{ cm})(-20.0 \text{ cm})/[20.0 \text{ cm} - (-20.0 \text{ cm})] = -10.0 \text{ cm}$$

and the magnification is $M = -(-10.0 \text{ cm})/(20.0 \text{ cm}) = +1/2$.

Thus, the image is virtual, located 10.0 cm in front of the lens, upright, and one-half the size of the object.   ◊

(c) When $p = +10.0$ cm, the image distance is

$$q = (10.0 \text{ cm})(-20.0 \text{ cm})/[10.0 \text{ cm} - (-20.0 \text{ cm})] = -6.67 \text{ cm}$$

and the magnification is $M = -(-6.67 \text{ cm})/(10.0 \text{ cm}) = +2/3$.

Thus, the image is virtual, located 6.67 cm in front of the lens, upright, and two-thirds the size of the object.   ◊

39. A converging lens is placed 30.0 cm to the right of a diverging lens of focal length 10.0 cm. A beam of parallel light enters the diverging lens from the left, and the beam is again parallel when it emerges from the converging lens. Calculate the focal length of the converging lens.

Solution

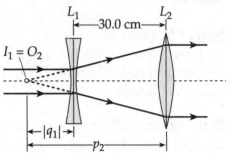

When parallel rays enter the divergent lens (L_1), it causes those rays to diverge as though they originated at the focal point on the front side of the lens. This means that it forms a virtual image with image distance $|q_1| = |f_1|$.

We can also show this by use of the thin lens equation. For parallel incident rays, $p_1 \to \infty$, and $1/p_1 \to 0$. Thus, the thin lens equation becomes

$$0 + \frac{1}{q_1} = \frac{1}{f_1} \quad \text{giving} \quad q_1 = f_1 = -10.0 \text{ cm}$$

The virtual image I_1 formed by the divergent lens is located in front of the convergent lens L_2. Therefore, this image serves as a real object for the convergent lens, with an object distance of

$$p_1 = +\left(|q_1| + 30.0 \text{ cm}\right) = +\left(10.0 \text{ cm} + 30.0 \text{ cm}\right) = +40.0 \text{ cm}$$

If the rays emerging from the convergent lens are to be parallel, the object for this lens must be at the focal point located in front of the lens. This means that the focal length of this lens is

$$f_2 = p_2 = +40.0 \text{ cm} \qquad \qquad \Diamond$$

This can also be seen from the thin lens equation. For parallel emerging rays, $q_2 \to \infty$, and $1/q_2 \to 0$. The thin lens equation then gives

$$\frac{1}{p_2} + 0 = \frac{1}{f_2} \quad \text{or} \quad f_2 = p_2 = +40.0 \text{ cm} \qquad \qquad \Diamond$$

45. Lens L_1 in Figure P23.45 has a focal length of 15.0 cm and is located a fixed distance in front of the film plane of a camera. Lens L_2 has a focal length of 13.0 cm, and its distance d from the film plane can be varied from 5.00 cm to 10.0 cm. Determine the range of distances for which objects can be focused on the film.

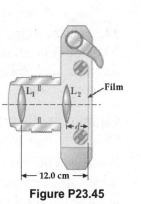

Figure P23.45

Solution

<u>Note:</u> Final answers to this problem are highly sensitive to round-off error. To avoid this, we retain extra digits in some intermediate answers and round only the final answers to the correct number of significant figures.

Both lenses are convergent lenses, so their focal lengths are $f_1 = +15.0$ cm and $f_2 = +13.0$ cm. The image formed by lens L_1 serves as the object for lens L_2. Since L_2 forms a real image on the film, the image distance for this lens is $q_2 = +d$, and the thin lens equation gives its object distance as

$$p_2 = \frac{q_2 f_2}{q_2 - f_2} = \frac{d(13.0 \text{ cm})}{d - 13.0 \text{ cm}}$$

Since the distance d is less than 12.0 cm (see Figure P23.45), the above equation shows that $p_2 < 0$. Thus, the image lens L_1 tries to form is located to the right of L_2 and serves as a virtual object for that lens. Since L_1 is a fixed distance of 12.0 cm from the film, the image distance for L_1 is

$$q_1 = 12.0 \text{ cm} + \left(|p_2| - d\right)$$

When lens L_2 is distance $d = 5.00$ cm in front of the film, then

$$p_2 = \frac{(5.00 \text{ cm})(13.0 \text{ cm})}{5.00 \text{ cm} - 13.0 \text{ cm}} = -8.125 \text{ cm, and}$$

$$q_1 = 12.0 \text{ cm} + (8.125 \text{ cm} - 5.00 \text{ cm}) = 15.125 \text{ cm}$$

The object distance for L_1 in this camera setting is

$$p_1 = \frac{q_1 f_1}{q_1 - f_1} = \frac{(15.125 \text{ cm})(15.0 \text{ cm})}{15.125 \text{ cm} - 15.0 \text{ cm}} = 1820 \text{ cm} = 18.2 \text{ m} \qquad \lozenge$$

When lens L_2 is distance $d = 10.0$ cm in front of the film, then

$$p_2 = \frac{(10.0 \text{ cm})(13.0 \text{ cm})}{10.0 \text{ cm} - 13.0 \text{ cm}} = -43.3 \text{ cm, and}$$

$$q_1 = 12.0 \text{ cm} + (43.3 \text{ cm} - 10.0 \text{ cm}) = 45.3 \text{ cm}$$

The object distance for L_1 in this camera setting is

$$p_1 = \frac{q_1 f_1}{q_1 - f_1} = \frac{(45.3 \text{ cm})(15.0 \text{ cm})}{45.3 \text{ cm} - 15.0 \text{ cm}} = 22.4 \text{ cm} = 0.224 \text{ m}$$ ◊

The camera can focus images on the film of objects located anywhere between 0.224 m and 18.2 m in front of the camera. ◊

---

52. The object in Figure P23.52 is midway between the lens and the mirror. The mirror's radius of curvature is 20.0 cm, and the lens has a focal length of –16.7 cm. Considering only the light that leaves the object and travels first towards the mirror, locate the final image formed by this system. Is the image real or virtual? Is it upright or inverted? What is the overall magnification of the image?

Solution

The ray diagram given below (*not to scale*) shows the formation of the final image. Rays leaving the original object O are caused to converge by the mirror. These rays would form a real image I_m if allowed to do so. However, the lens intercepts and redirects these rays, causing them to diverge as though they came from the upright, virtual, final image I_L.

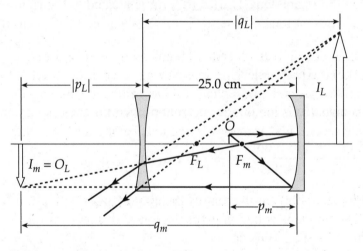

The object distance for the mirror is $p_m = (25.0 \text{ cm})/2 = +12.5 \text{ cm}$

The mirror equation then gives the image distance for the mirror as

$$q_m = \frac{p_m R}{2 p_m - R} = \frac{(12.5 \text{ cm})(20.0 \text{ cm})}{2(12.5 \text{ cm}) - 20.0 \text{ cm}} = +50.0 \text{ cm}$$

This real image that the mirror would form serves as the object for the lens. Since it is located behind the lens (that is, on the side opposite the incident light), it is a virtual object, and the object distance for the lens is negative. This object distance is

$$p_L = -(q_m - 25.0 \text{ cm}) = -(50.0 \text{ cm} - 25.0 \text{ cm}) = -25.0 \text{ cm}$$

The image distance for the lens is

$$q_L = \frac{p_L f_L}{p_L - f_L} = \frac{(-25.0 \text{ cm})(-16.7 \text{ cm})}{-25.0 \text{ cm} - (-16.7 \text{ cm})} = -50.3 \text{ cm}$$

Since $q_L < 0$, the final image is a virtual image, located 50.3 cm in front of the lens (that is, on the same side as the incident light) or 25.3 cm behind the mirror. ◊

The overall magnification produced by this mirror-lens system is given by

$$M_{overall} = M_m M_L$$

where

$$M_m = -\frac{q_m}{p_m} = -\frac{50.0 \text{ cm}}{12.5 \text{ cm}} = -4.00$$

is the magnification by the mirror, and

$$M_L = -\frac{q_L}{p_L} = -\frac{-50.3 \text{ cm}}{-25.0 \text{ cm}} = -2.01$$

is the magnification by the lens. Thus, $M_{overall} = (-4.00)(-2.01) = +8.04$ ◊

Since $M_{overall} > 0$, the final image is upright and 8.04 times the size of the original object. ◊

---

57. An object 2.00 cm high is placed 40.0 cm to the left of a converging lens having a focal length of 30.0 cm. A diverging lens having a focal length of −20.0 cm is placed 110 cm to the right of the converging lens. (a) Determine the final position and magnification of the final image. (b) Is the image upright or inverted? (c) Repeat parts (a) and (b) for the case where the second lens is a converging lens having a focal length of +20.0 cm.

Solution

With the object located 40.0 cm in front of the converging first lens, $p_1 = +40.0$ cm and $f_1 = +30.0$ cm. The thin lens equation gives the image distance for this lens as

$$q_1 = \frac{p_1 f_1}{p_1 - f_1} = \frac{(40.0 \text{ cm})(30.0 \text{ cm})}{40.0 \text{ cm} - 30.0 \text{ cm}} = +120 \text{ cm}$$

and the magnification due to this lens is $M_1 = -\frac{q_1}{p_1} = -\frac{120 \text{ cm}}{40.0 \text{ cm}} = -3.00$

The real image the first lens tries to form 120 cm beyond that lens serves as the object for the second lens. Since the two lenses are only 110 cm apart, the location of this object is to the right of the second lens as shown in the diagram below. Thus, this is a virtual object and the object distance for the second lens is $p_2 = 110 \text{ cm} - q_1 = -10.0 \text{ cm}$.

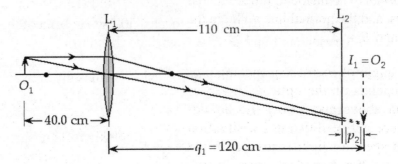

(a) If $f_2 = -20.0$ cm, the image distance and magnification for the second lens, and the overall magnification is

$$q_2 = \frac{p_2 f_2}{p_2 - f_2} = \frac{(-10.0 \text{ cm})(-20.0 \text{ cm})}{-10.0 \text{ cm} - (-20.0 \text{ cm})} = +20.0 \text{ cm} \qquad \lozenge$$

$$M_2 = -\frac{q_2}{p_2} = -\frac{+20.0 \text{ cm}}{-10.0 \text{ cm}} = +2.00 \, ; \quad M = M_1 M_2 = (-3.00)(+2.00) = -6.00 \qquad \lozenge$$

(b) The final image is real ($q_2 > 0$), inverted ($M < 0$), located 20.0 cm to the right of the second lens, and $|h'| = h|M| = (2.00 \text{ cm})(6.00) = 12.0$ cm high. $\qquad \lozenge$

(c) If $f_2 = +20.0$ cm, the image distance and magnification for the second lens, and the overall magnification is

$$q_2 = \frac{p_2 f_2}{p_2 - f_2} = \frac{(-10.0 \text{ cm})(20.0 \text{ cm})}{-10.0 \text{ cm} - 20.0 \text{ cm}} = +6.67 \text{ cm} \qquad \lozenge$$

$$M_2 = -\frac{q_2}{p_2} = -\frac{+6.67 \text{ cm}}{-10.0 \text{ cm}} = +0.667 \, ; \quad M = M_1 M_2 = (-3.00)(+0.667) = -2.00 \qquad \lozenge$$

The final image is real ($q_2 > 0$), inverted ($M < 0$), located 6.67 cm beyond the second lens, and $|h'| = h|M| = (2.00 \text{ cm})(2.00) = 4.00$ cm high. $\qquad \lozenge$

61. The lens maker's equation for a lens with index n_1 immersed in a medium with index n_2 takes the form

$$\frac{1}{f} = \left(\frac{n_1}{n_2} - 1\right)\left(\frac{1}{R_1} - \frac{1}{R_2}\right)$$

A thin diverging glass (index = 1.50) lens with $R_1 = -3.00$ m and $R_2 = -6.00$ m is surrounded by air. An arrow is placed 10.0 m to the left of the lens. (a) Determine the position of the image. Repeat part (a) with the arrow and lens immersed in (b) water (index = 1.33); (c) a medium with an index of refraction of 2.00. (d) How can a lens that is diverging in air be changed into a converging lens?

Solution

For the lens described above, the lens maker's equation gives

$$\frac{1}{f} = \left(\frac{1.50}{n_2} - 1\right)\left(\frac{1}{-3.00 \text{ m}} + \frac{1}{6.00 \text{ m}}\right) = \left(\frac{1.50 - n_2}{n_2}\right)\left(\frac{-1}{6.00 \text{ m}}\right)$$

or $f = \dfrac{n_2(6.00 \text{ m})}{n_2 - 1.50}$ (1)

(a) If $n_2 = n_{\text{air}} = 1.00$, then $f = \dfrac{1.00(6.00 \text{ m})}{1.00 - 1.50} = -12.0$ m

The thin lens equation gives

$$q = \frac{pf}{p - f} = \frac{(10.0 \text{ m})(-12.0 \text{ m})}{10.0 \text{ m} - (-12.0 \text{ m})} = -5.45 \text{ m}$$

So the lens forms a virtual image 5.45 m to the left of the lens.   ◊

(b) If $n_2 = n_{\text{water}} = 1.33$, then $f = \dfrac{1.33(6.00 \text{ m})}{1.33 - 1.50} = -46.9$ m

The thin lens equation gives

$$q = \frac{pf}{p - f} = \frac{(10.0 \text{ m})(-46.9 \text{ m})}{10.0 \text{ m} - (-46.9 \text{ m})} = -8.24 \text{ m}$$

and the lens forms a virtual image 8.24 m to the left of the lens.   ◊

(c) If $n_2 = 2.00$, then $f = \dfrac{2.00(6.00 \text{ m})}{2.00 - 1.50} = +24.0 \text{ m}$

The thin lens equation gives

$$q = \frac{pf}{p-f} = \frac{(10.0 \text{ m})(24.0 \text{ m})}{10.0 \text{ m} - 24.0 \text{ m}} = -17.1 \text{ m}$$

and the lens forms a virtual image 17.1 m to the left of the lens.   ◊

(d) From Equation (1), we see that this lens (which is diverging in air) will have a positive focal length and act as a converging lens when $n_2 > 1.50$. Thus, a lens which is diverging in air can be changed into a converging lens by surrounding it with a medium having a refractive index greater than that of the lens material.   ◊

24

Wave Optics

NOTES FROM SELECTED CHAPTER SECTIONS

24.1 Conditions for Interference

In order to observe **sustained interference** in light waves, the following conditions must be met:

- The **sources must be coherent;** they must maintain a constant phase with respect to each other.

- The **sources must be monochromatic;** they must have identical wavelengths.

- The **superposition principle must apply.**

24.2 Young's Double-Slit Experiment

The figure at right is a schematic diagram (not to scale) of the experimental setup of Young's double-slit experiment. The source S_0 is a monochromatic source that is divided into two coherent sources by passing through slits S_1 and S_2. The resulting interference pattern is observed on a viewing screen as bright and dark fringes corresponding to constructive and destructive interference of light from slits S_1 and S_2.

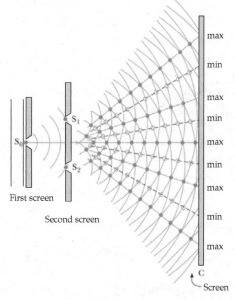

Constructive interference (a bright fringe) occurs when the path difference equals either zero or an integral multiple of the wavelength.

Destructive interference (a dark fringe) results when the path difference equals an odd multiple of a half wavelength.

24.3   Change of Phase Due to Reflection

Consider a light wave traveling in a medium with an index of refraction n_1 and experiencing partial reflection at the surface of a medium with an index of refraction n_2 as shown in the figure below.

- At Surface A, where $n_1 < n_2$, the reflected ray experiences a phase change of 180°.

- At Surface B, where $n_1 > n_2$, the reflected ray experiences no phase change.

- There is no phase change in the transmitted ray regardless of the relative values of n_1 and n_2.

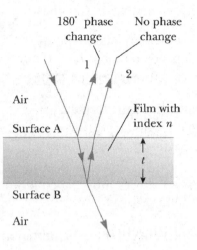

Phase change upon reflection

24.4   Interference in Thin Films

In order to predict constructive or destructive interference in thin films you must consider:

- the difference in path-length traveled by the two interfering waves (reflected from the top and bottom surfaces).

- any expected changes in phase due to reflection.

- the wavelength of the light within the film.

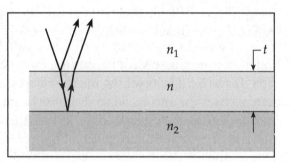

There are two general cases to consider (see figure above):

(a)   **Reflection resulting in a phase change at only one surface of the film:**

n_1 and n_2 both less than n (phase change occurs only at the top surface), or

n_1 and n_2 both greater than n (phase change occurs only at the bottom surface)

Constructive interference will occur under these conditions when the path difference ($2t$) is an odd number of half wavelengths *as measured in the film* (i.e., $t = (m + \frac{1}{2})\lambda / 2n$, where λ / n is the wavelength in the film).

Destructive interference will occur under these conditions when the path difference ($2t$) equals an integral number of wavelengths *as measured in the film* (i.e., $t = m\lambda / 2n$).

(b) Reflection resulting in phase changes at both top and bottom surfaces of the film (or at neither surface):

$n_1 < n$ and $n < n_2$ (phase change at both surfaces), or

$n_1 > n$ and $n > n_2$ (no phase change at either surface)

In this case, any phase changes that occur are offsetting and interference of the reflected rays depends only on the difference in distance traveled by the two reflected rays and the index of refraction of the film.

Constructive interference will occur when the path difference ($2t$) equals an integral number of wavelengths *as measured in the film*; the film thickness must be an integral number of half wavelengths; that is, $t = m\lambda / 2n$.

Destructive interference in this case will be observed when the path difference equals an odd number of half wavelengths *as measured in the film*; that is, when $t = \left(m + \frac{1}{2}\right)\lambda / 2n$.

24.6 Diffraction

24.7 Single-Slit Diffraction

Diffraction occurs when light waves deviate (or spread) from their initial direction of travel when passing through small openings, around obstacles, or by sharp edges.

The diffraction pattern produced by a single narrow slit consists of a broad, intense central band (the central maximum), flanked by a series of narrower and less intense secondary bands (called secondary maxima) alternating with a series of dark bands, or minima.

In the case of **single slit diffraction,** each portion of the slit acts as a source of waves; and light from one portion of the slit can interfere with light from another portion. The resultant intensity on the screen depends on angle θ, which determines the direction between the perpendicular to the plane of the slit and the direction to a point on the screen.

One type of diffraction, called **Fraunhofer diffraction,** occurs when the rays reaching the observing screen are approximately parallel.

24.8 The Diffraction Grating

A diffraction grating, consisting of many equally spaced parallel slits, separated by a distance d, will produce a diffraction pattern. *There will be a series of principal maxima (bright lines) for each wavelength component in the incident light.*

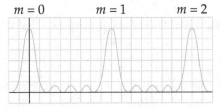

The zeroth-, first-, and second-order principal maxima of a single-wavelength diffraction pattern.

The figure illustrates the case for which the incident light contains a single wavelength component. Although not shown in the figure above, there will also be maxima corresponding to $m = -1, -2, \ldots$. These maxima occur at angles θ, measured from the line perpendicular to the grating, where $m\lambda = d\sin\theta$.

If the incident light contains a second wavelength component, a second series of principal maxima (in general with a different intensity) will be present in the diffraction pattern.

In a given spectral order, denoted by the number *m*, there will be principal maxima corresponding to each wavelength component incident on the grating.

24.9   Polarization of Light Waves

The electric field vector of a light wave vibrates in a plane perpendicular to the direction of propagation. In the figure the direction of propagation is out of the page. As illustrated in the top figure, for **unpolarized light**, the electric field vector vibrates along all directions in the plane perpendicular to the direction of travel with equal probability. However, at a given point and at a particular instant, there is only one resultant electric field direction (shown by the electric field vector $\vec{E}$).

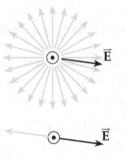

In **linearly polarized light** the electric field vibrates along the same direction at all times at a particular point, as illustrated in the bottom figure.

It is possible to obtain a linearly polarized wave from an unpolarized wave by removing from the unpolarized wave all components except those whose electric field vectors oscillate in a single plane. The plane formed by the direction of vibration of the electric field and the direction of propagation is called the **plane of polarization.**

Three important processes for producing linearly polarized light are: (1) selective absorption, (2) reflection, and (3) scattering.

Optical activity is the property of certain materials to cause rotation of the plane of polarization as a light beam travels through the material.

EQUATIONS AND CONCEPTS

In **Young's double-slit experiment,** two slits, S_1 and S_2, separated by a distance *d*, serve as monochromatic coherent sources. The light intensity at any point on the screen is the resultant of light reaching the screen from both slits. As illustrated in the figure, a point *P* on the screen can be identified by the angle θ or by the distance *y* from the center of the screen: $y = L \tan\theta \approx L \sin\theta$.

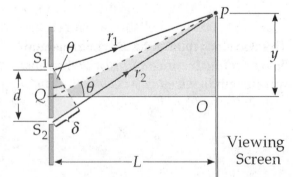

Geometry of Young's Double-Slit Experiment

A **path difference** δ arises because waves from S_1 and S_2 travel unequal distances to reach a point on the screen (except the center point, O). *The value of δ determines whether the waves from the two slits arrive in phase or out of phase.*

$$\delta = r_2 - r_1 = d\sin\theta \qquad (24.1)$$

Constructive interference (bright fringes) will appear at points on the screen for which the path difference is equal to an integral multiple of the wavelength. The positions of bright fringes can also be located by calculating their vertical distance from the center of the screen (y). *In each case, the number m is called the order number of the fringe.* The central bright fringe ($\theta = 0°$, $m = 0$) is called the **zeroth-order maximum** and the first maximum on either side of the central maximum is called the **first order maximum.** Equations (24.5) and (24.6) assume that the angle θ in the figure above is small (less than 5°).

Conditions for double-slit constructive interference:

$$\delta = d\sin\theta_{\text{bright}} = m\lambda \qquad (24.2)$$

$$m = 0, \ \pm 1, \ \pm 2, \ldots$$

$$y_{\text{bright}} = \left(\frac{\lambda L}{d}\right) m \qquad (24.5)$$

$$m = 0, \ \pm 1, \ \pm 2, \ldots$$

m is called the order number

Dark fringes (destructive interference) will appear at points on the screen which correspond to path differences of an odd multiple of half wavelengths. For these points of destructive interference, waves which leave the two slits in phase arrive at the screen 180° (one-half wavelength) out of phase.

Conditions for double-slit destructive interference:

$$\delta = d\sin\theta_{\text{dark}} = \left(m + \tfrac{1}{2}\right)\lambda \qquad (24.3)$$

$$m = 0, \ \pm 1, \ \pm 2, \ldots$$

$$y_{\text{dark}} = \left(\frac{\lambda L}{d}\right)\left(m + \tfrac{1}{2}\right) \qquad (24.6)$$

$$m = 0, \ \pm 1, \ \pm 2, \ldots$$

The **wavelength of light,** λ_n, in a medium having refractive index n, is less than the wavelength in vacuum, λ.

$$\lambda_n = \frac{\lambda}{n} \qquad (24.7)$$

Interference in thin films depends on wavelength, film thickness, and the indices of refraction of the film and surrounding media. *Differences in phase may be due to path difference, phase change upon reflection, or both.* There are two general cases as described below.

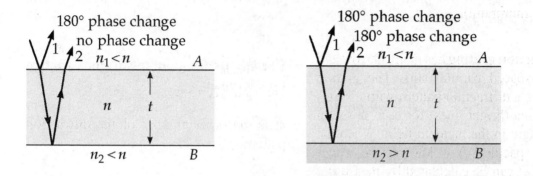

Case (I)   Phase change at only one film surface

Indices of refraction of media on both sides of the film are less than that of the film ($n_1 < n$ and $n_2 < n$) as shown in the figure above left; or indices on both both sides are greater than that of the film ($n_1 > n$ and $n_2 > n$).

Constructive interference
$$2nt = \left(m + \tfrac{1}{2}\right)\lambda \qquad (24.9)$$
($m = 0, 1, 2, \ldots$)

Destructive interference
$$2nt = m\lambda \qquad (24.10)$$
($m = 0, 1, 2, \ldots$)

Case (II)   Phase changes at both or neither surface

Film is between two media either of which has an index of refraction greater than that of the film and the other a smaller index ($n_1 < n < n_2$) as shown in the figure above right; or ($n_1 > n > n_2$).

Constructive interference
$$2nt = m\lambda$$
($m = 0, 1, 2, \ldots$)

Destructive interference
$$2nt = \left(m + \tfrac{1}{2}\right)\lambda$$
($m = 0, 1, 2, \ldots$)

Note that the roles of the equations in Case (I) and Case (II) are reversed for constructive and destructive interference.

In **single-slit diffraction,** the total phase difference between waves from the top and bottom portions the slit will depend on the angle θ which determines the direction to a point on the screen. The pattern consists of a broad central bright band and a series of less intense and narrower side bands.

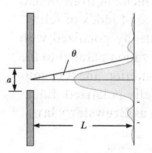

Single slit diffraction pattern

The **general condition for destructive interference** (dark band) at a given point on the screen can be stated in terms of the angle θ_{dark}. Equation (24.11) gives the angle θ_{dark} for which the intensity will be zero (or minimum).

For destructive interference:

$$\sin\theta_{dark} = m\frac{\lambda}{a} \qquad (24.11)$$

$$m = \pm 1, \ \pm 2, \ \pm 3, \ldots$$

A **diffraction grating** has many narrow and equally-spaced parallel slits. The grating produces a diffraction pattern with a series of maxima (bright lines) for each different wavelength in the incident light. When the grating spacing (d) is known, the wavelength (λ) can be calculated by measuring the angle (θ_{bright}) for a given spectral line. Maxima due to wavelengths of different values comprise a spectral order denoted by an **order number** (m).

$$d\sin\theta_{bright} = m\lambda \qquad (24.12)$$

$$(m = \pm 1, \ \pm 2, \ \pm 3 \pm \ldots)$$

m is the order number of the diffraction pattern.

Malus's law states that the fraction of initially polarized light (from a polarizer) that will be transmitted by a second sheet of polarizing material (the analyzer) depends on the square of the cosine of the angle θ between the transmission axis of the polarizer and that of the analyzer. The transmitted intensity is a maximum when the transmission axes are parallel ($\theta = 0°$ or $180°$). When the transmission axes are perpendicular to each other, the light is completely absorbed by the analyzer, and the transmitted intensity is zero.

$$I = I_0 \cos^2\theta \qquad (24.13)$$

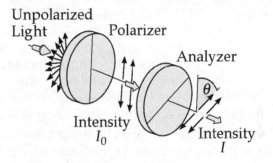

I_0 is the intensity of the beam incident on the analyzer after being transmitted by the polarizer.

The **polarizing angle** is the angle of incidence for which light incident from air and reflected from a surface of index of refraction, n, will be completely polarized with the direction of polarization parallel to the surface. Under this condition, the transmitted light will be partially polarized. Equation (24.14) is known as **Brewster's law.**

$$n = \tan\theta_p \qquad (24.14)$$

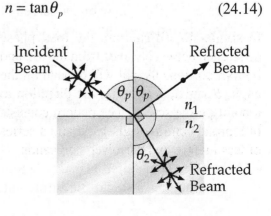

SUGGESTIONS, SKILLS, AND STRATEGIES

THIN FILM INTERFERENCE PROBLEMS

1. Identify the thin film from which interference effects are being observed.

2. The type of interference (constructive or destructive) that occurs in a specific situation is determined by the phase relationship between that portion of the wave reflected at the upper surface of the film and that portion reflected at the lower surface of the film.

3. The phase difference between the two portions of the reflected wave is determined by: (i) The physical path difference and (ii) Any phase changes which may occur at either surface upon reflection. *You must consider both potential contributions to the total phase difference.*

    Path difference – The wave reflected from the lower surface of the film has to travel a distance equal to **twice the thickness of the film** before it returns to the upper surface of the film where it interferes with that portion of the wave reflected at the upper surface.

    Phase change upon reflection – In addition to path difference, reflections may change the predicted interference results. When a wave traveling in a particular medium reflects off a surface having a higher index of refraction than the one in which it was initially traveling, a 180° phase shift occurs. This has the same effect as if the wave lost $\frac{1}{2}\lambda$. Phase changes due to reflection must be considered in addition to the extra distance traveled by the reflected wave.

4. **When path difference and phase changes upon reflection are both taken into account,** the interference will be constructive if the two waves are out of phase ("out of step") by an integral multiple of λ_n (the wavelength as measured in the film). Destructive interference will occur when they differ in phase by an odd number of half-wavelengths.

REVIEW CHECKLIST

*   Describe Young's double-slit experiment to demonstrate the wave nature of light. Account for the phase difference between light waves from the two sources as they arrive at a given point on the screen. State the conditions for constructive and destructive interference in terms of each of the following: path difference (δ), distance from center of screen (y), and the angle between the perpendicular bisector of the two sources and the location on the screen (θ). (Section 24.2)

*   Account for the conditions of constructive and destructive interference in thin films considering both path difference and any expected phase changes due to reflection. (Sections 24.3 and 24.4)

*   Describe Fraunhofer diffraction produced by a single slit. Determine the positions of the minima in a single-slit diffraction pattern. (Sections 24.6 and 24.7)

- Determine the positions of the principal maxima (or calculate wavelength values) in spectra formed by a diffraction grating. (Section 24.8)

- Describe qualitatively the polarization of light by selective absorption, reflection, and scattering. Also, make appropriate calculations using Brewster's law and Malus's law. (Section 24.9)

SOLUTIONS TO SELECTED END-OF-CHAPTER PROBLEMS

3. A pair of narrow, parallel slits separated by 0.250 mm is illuminated by the green component from a mercury vapor lamp ($\lambda = 546.1$ nm). The interference pattern is observed on a screen 1.20 m from the plane of the parallel slits. Calculate the distance (a) from the central maximum to the first bright region on either side of the central maximum and (b) between the first and second dark bands in the interference pattern.

Solution

When waves pass through a pair of parallel slits, the slits act as a set of coherent sources. From the sketch at the right, observe that the difference in path lengths is $\delta = d \sin\theta$, where d is the spacing between slits and θ is measured from the line perpendicular to the plane of the slits.

(a) To have constructive interference, the path difference δ must be an integral number of wavelengths. Thus, bright fringes occur at angles θ_m where

$$\delta = d \sin\theta_m = m\lambda \qquad m = 0, \pm 1, \pm 2, \pm 3, \ldots$$

In most cases, with slits illuminated by visible light, the angles θ_m are very small and $\sin\theta_m \approx \tan\theta_m$. Then, the distance from the central maximum to the bright fringe of order m on a screen located distance L from the slits is given by

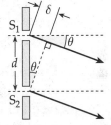

$$y_m = L\tan\theta_m \approx L\sin\theta_m = m\left(\frac{\lambda L}{d}\right)$$

Since $m = 0$ for the central maximum and $m = \pm 1$ for the first bright fringe on either side of the central maximum, the distance from the central maximum to the first bright fringe is

$$y_1 = (1)\frac{\lambda L}{d} = \frac{\left(546.1 \times 10^{-9}\ \text{m}\right)(1.20\ \text{m})}{0.250 \times 10^{-3}\ \text{m}} = 2.62 \times 10^{-3}\ \text{m} = 2.62\ \text{mm} \qquad \lozenge$$

(b) For destructive interference and a dark fringe, the path difference δ must be an odd number of half-wavelengths. Thus, dark fringes occur where

$$\delta = d \sin\theta_m = \left(m + \tfrac{1}{2}\right)\lambda \ \text{ and } \ y_m = \left(m + \tfrac{1}{2}\right)\frac{\lambda L}{d} \quad m = 0, \pm 1, \pm 2, \pm 3, \ldots$$

The spacing between the first ($m = 0$) and second ($m = 1$) dark fringes is then

$$\Delta y = y_1 - y_0 = \frac{3\,\lambda L}{2\,d} - \frac{\lambda L}{2\,d} = \frac{\lambda L}{d} = \frac{\left(546.1 \times 10^{-9}\ \text{m}\right)\left(1.20\ \text{m}\right)}{0.250 \times 10^{-3}\ \text{m}} = 2.62\ \text{mm} \qquad \Diamond$$

---

13. Radio waves from a star, of wavelength 250 m, reach a radio telescope by two separate paths as shown in Figure P24.13. One is a direct path to the receiver, which is situated on the edge of a cliff by the ocean. The second is by reflection off the water. The first minimum of destructive interference occurs when the star is 25.0° above the horizon. Find the height of the cliff. (Assume no phase change on reflection.)

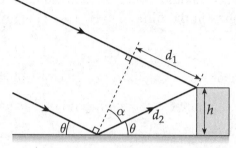

Modified Figure P24.13

Solution

If the star is at angle θ above the horizon, the incident rays will be at angle θ above the horizontal as shown in the sketch. The law of reflection then requires that the reflected ray travel at angle θ above the horizontal. Observe from the sketch that the difference in the distances the direct ray and the reflected ray have traveled when they arrive at the receiver is

$$\delta = d_2 - d_1 = d_2 - d_2 \sin\alpha = d_2\left(1 - \sin\alpha\right)$$

From the sketch, note that $\qquad \theta + 90° + \alpha + \theta = 180° \qquad$ or $\qquad \alpha = 90° - 2\theta$

Thus, when $\theta = 25.0°$, we have $\alpha = 40.0°$ and $\delta = d_2\left(1 - \sin 40.0°\right)$

If we assume no phase shift on reflection, this difference in path lengths must equal $\lambda/2 = 125$ m when the first minimum of intensity caused by destructive interference occurs. Hence, we now have

$$125\ \text{m} = d_2\left(1 - \sin 40.0°\right) \qquad \text{or} \qquad d_2 = \frac{125\ \text{m}}{1 - \sin 40.0°} = 350\ \text{m}$$

Looking at the sketch again, we see that the height of the cliff is

$$h = d_2 \sin\theta = \left(350\ \text{m}\right)\sin 25.0° = 148\ \text{m} \qquad \Diamond$$

---

21. A possible means for making an airplane invisible to radar is to coat the plane with an antireflective polymer. If radar waves have a wavelength of 3.00 cm, and the index of refraction of the polymer is $n = 1.50$, how thick would you make the coating?

Solution

The concept of an antireflective coating is to produce destructive interference in the waves reflecting from the first and second surfaces of the coating. Assuming the refractive index of the polymer is less than that of the material making up the body of the airplane, waves reflecting from both surfaces of the coating will experience phase reversals, or 180° phase shifts. The reversal at one surface offsets the reversal at the other surface, so the net path difference in the two reflected waves is $\delta = 2t$ where t is the thickness of the film. For destructive interference, this path difference should equal an odd number of half wavelengths in the film, $\lambda_n = \lambda/n_{film}$. Thus, we require that

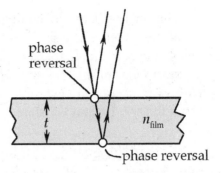

$$2t = \left(m + \frac{1}{2}\right)\lambda_n = \left(m + \frac{1}{2}\right)\frac{\lambda}{n_{film}} \qquad \text{where} \qquad m = 0, 1, 2, \ldots$$

For the thinnest coating meeting this condition, $m = 0$. Then, the required thickness of the coating is

$$t = \frac{\lambda}{4n_{film}} = \frac{3.00 \text{ cm}}{4(1.50)} = 0.500 \text{ cm} \qquad\qquad \lozenge$$

Basing your defensive strategy on a coating such as this would be very risky. The enemy could modify its radar to use a wavelength of 1.50 cm. Then, your polymer coating would make the airplanes very good reflectors by producing constructive interference in the reflected radar waves.

---

27. A planoconvex lens rests with its curved side on a flat glass surface and is illuminated from above by light of wavelength 500 nm. (See Figure 24.8.) A dark spot is observed at the center, surrounded by 19 concentric dark rings (with bright rings in between). How much thicker is the air wedge at the position of the 19th dark ring than at the center?

Solution

When light reflects while trying to cross a boundary going from one medium into a second medium with a higher index of refraction, it experiences a phase reversal or a 180° shift in phase. If the reflection occurs as the light is attempting to go from one medium into one of lower index of refraction, no such phase reversal occurs.

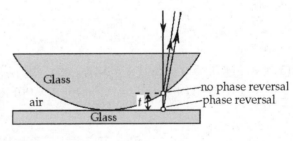

As shown in the sketch above, the planoconvex lens and glass plate have an air film of varying thickness between them. The light reflecting at the upper surface of this film does not experience a phase reversal, while light reflecting at the lower surface of the film does undergo a phase reversal. With only one phase reversal involved with the two reflections, Equation 24.10 in the textbook gives the condition for destructive interference in the reflected light as

$$2nt = m\lambda \qquad\qquad m = 0, 1, 2, 3, \ldots$$

where n is the index of refraction of the film (air in this case), t is the thickness of the film at the point of reflection, and λ is the wavelength of the light.

Note that $m = 0$ corresponds to a thickness of $t = 0$. Hence, there will be a dark spot in the interference pattern at the point of contact between the lens and flat plate. Higher values of m correspond to dark fringes in the interference pattern. In this case, the fringes will be circular in shape, centered on the central dark spot, located where the air film has the thickness associated with that value of m in the above expression. The 19th dark ring occurs for $m = 19$ and is located where the film has thickness

$$t = \frac{m\lambda}{2n_{air}} = \frac{(19)(500 \times 10^{-9} \text{ m})}{2(1.00)} = 4.75 \times 10^{-6} \text{ m} = 4.75 \ \mu\text{m}$$ ◊

---

33. Light of wavelength 587.5 nm illuminates a single slit of width 0.75 mm. (a) At what distance from the slit should a screen be placed if the first minimum in the diffraction pattern is to be 0.85 mm from the central maximum? (b) Calculate the width of the central maximum.

Solution

Light passing through different portions of a single slit interferes in such a way as to produce maxima and minima in intensity on a screen located at distance L beyond the slit as shown in the sketch at the right. The minima or dark fringes occur where

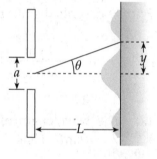

$$\sin\theta = m\lambda/a \quad \text{with} \quad m = \pm1, \pm2, \pm3, \ldots$$

When the slit width, a, is very large in comparison to the wavelength, λ, of the light, the angles θ are very small. In this case, $\sin\theta \approx \tan\theta = y/L$ so the locations of the minima on the screen are given by

$$y_m = mL\left(\frac{\lambda}{a}\right) \quad \text{with} \quad m = \pm1, \pm2, \pm3, \ldots$$

(a) If $a = 0.75$ mm and $\lambda = 587.5$ nm, the slit to screen distance required to produce the first $(m = 1)$ minimum at 0.85 mm from the central maximum is

$$L = \frac{y_1}{1}\left(\frac{a}{\lambda}\right) = (0.85 \times 10^{-3} \text{ m})\left(\frac{0.75 \times 10^{-3} \text{ m}}{587.5 \times 10^{-9} \text{ m}}\right) = 1.1 \text{ m}$$ ◊

(b) The central maximum extends from the first minimum on one side of the symmetry line to the first minimum on the other side of this line. Thus, its width on the screen is

$$\Delta y = y_1 - y_{-1} = 2y_1 = 2(0.85 \text{ mm}) = 1.7 \text{ mm}$$ ◊

39. Three discrete spectral lines occur at angles of 10.1°, 13.7°, and 14.8°, respectively, in the first-order spectrum of a diffraction-grating spectrometer. (a) If the grating has 3 660 slits/cm, what are the wavelengths of the light? (b) At what angles are these lines found in the second-order spectra?

Solution

When light diffracts through a grating, constructive inter-ference occurs and bright lines are found when the path difference for light coming through two adjacent slits of the grating is an integer multiple of the wavelength of the light. The sketch at the right shows rays coming from two adjacent slits, separated by distance d. From this sketch, observe that the difference in path lengths for these rays is $\delta = d\sin\theta$ where θ is measured from the line perpendicular to the plane of the grating. The requirement for constructive interference, and formation of a bright line, is then

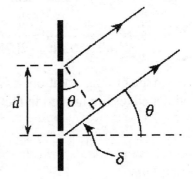

$$\delta = d\sin\theta = m\lambda \qquad \text{where} \qquad m = 0, 1, 2, 3, \ldots$$

For the described grating, the distance between adjacent slits is

$$d = \frac{1\text{ cm}}{3\,660\text{ slits}} = \frac{10^{-2}\text{ m}}{3\,660} = 2.73 \times 10^{-6}\text{ m} = 2\,732\text{ nm}$$

(a) The wavelength of light in a first order ($m = 1$) bright line found at angle θ is $\lambda = d\sin\theta$. Thus,

at $\theta = 10.1°$: $\lambda_1 = d\sin\theta_1 = (2\,732\text{ nm})\sin 10.1° = 479\text{ nm}$ ◊

at $\theta = 13.7°$: $\lambda_2 = d\sin\theta_2 = (2\,732\text{ nm})\sin 13.7° = 647\text{ nm}$ ◊

at $\theta = 14.8°$: $\lambda_3 = d\sin\theta_3 = (2\,732\text{ nm})\sin 14.8° = 698\text{ nm}$ ◊

(b) In the second order, $m = 2$ and the angle at which these wavelengths will form bright lines are:

for $\lambda_1 = 479\text{ nm}°$: $\theta = \sin^{-1}(2\lambda_1/d) = \sin^{-1}\left[2(479\text{ nm})/2\,732\text{ nm}\right] = 20.5°$ ◊

for $\lambda_2 = 647\text{ nm}°$: $\theta = \sin^{-1}(2\lambda_2/d) = \sin^{-1}\left[2(647\text{ nm})/2\,732\text{ nm}\right] = 28.3°$ ◊

for $\lambda_3 = 698\text{ nm}°$: $\theta = \sin^{-1}(2\lambda_3/d) = \sin^{-1}\left[2(698\text{ nm})/2\,732\text{ nm}\right] = 30.7°$ ◊

47. Sunlight is incident on a diffraction grating that has 2 750 lines/cm. The second-order spectrum over the visible range (400–700 nm) is to be limited to 1.75 cm along a screen that is a distance L from the grating. What is the required value of L?

Solution

The spacing between adjacent slits on this grating is

$$d = \frac{1 \text{ cm}}{2\,750 \text{ slits}} = \frac{10^{-2} \text{ m}}{2\,750 \text{ slits}} = 3.636 \times 10^{-6} \text{ m} = 3\,636 \text{ nm}$$

From the grating equation, $d \sin\theta = m\lambda$, where $m = 1, 2, 3, \ldots$, the angles where the red and violet edges are located in the second-order spectrum are:

$$\theta_r = \sin^{-1}\left(\frac{2\lambda_{red}}{d}\right) = \sin^{-1}\left[\frac{2(700 \text{ nm})}{3\,636 \text{ nm}}\right] = 22.65°$$

and $\quad \theta_v = \sin^{-1}\left(\frac{2\lambda_{violet}}{d}\right) = \sin^{-1}\left[\frac{2(400 \text{ nm})}{3\,636 \text{ nm}}\right] = 12.71°$

Observe from the sketch at the right, the screen positions (measured from the location of the central maximum) of the edges of this second-order spectrum are $y_r = L\tan\theta_r$ and $y_v = L\tan\theta_v$. Thus, the width of the second-order spectrum on the screen is

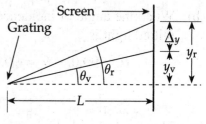

$$\Delta y = y_r - y_v = L\tan\theta_r - L\tan\theta_v = L(\tan\theta_r - \tan\theta_v)$$

If it is desired to have $\Delta y = 1.75$ cm as given, the distance between the grating and the screen must be

$$L = \frac{\Delta y}{\tan\theta_r - \tan\theta_v} = \frac{1.75 \text{ cm}}{\tan(22.65°) - \tan(12.71°)} = 9.13 \text{ cm}$$

---

51. The angle of incidence of a light beam in air onto a reflecting surface is continuously variable. The reflected ray is found to be completely polarized when the angle of incidence is 48.0°. (a) What is the index of refraction of the reflecting material? (b) If some of the incident light (at an angle of 48.0°) passes into the material below the surface, what is the angle of refraction?

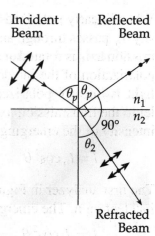

Solution

If, when light is incident on a surface, the reflected beam and the refracted beam are perpendicular to each other as shown in the figure at the right, the reflected beam will be completely polarized with the electric field vector parallel to the surface. The angle of incidence, θ_p, at which this occurs is called the polarizing angle or the Brewster's angle.

From the figure, observe that $\quad \theta_p + 90.0° + \theta_2 = 180°$

or $\quad\quad\quad\quad\quad\quad\quad\quad \theta_2 = 90.0° - \theta_p$

That is, when the angle of incidence is the polarizing angle, the angle of refraction is the complement of that angle, and using the trigonometric identity for the sine of the difference of two angles gives

$$\sin \theta_2 = \sin\left(90.0° - \theta_p\right) = \sin 90.0° \cos \theta_p - \cos 90.0° \sin \theta_p = \cos \theta_p$$

Thus, for light incident at the polarizing angle, Snell's law, $n_1 \sin\theta_1 = n_2 \sin\theta_2$, becomes

$$n_1 \sin\theta_p = n_2 \cos\theta_p \quad\quad \text{or} \quad\quad \frac{\sin\theta_p}{\cos\theta_p} = \tan\theta_p = \frac{n_2}{n_1}$$

(a) If the incident light is in air ($n_1 = 1.00$) and the polarizing angle is found to be $\theta_p = 48.0°$, the index of refraction of the reflecting material is

$$n_2 = n_1 \tan\theta_p = (1.00)\tan 48.0° = 1.11 \quad\quad\quad ◊$$

(b) When the angle of incidence is $\theta_1 = \theta_p = 48.0°$, the angle of refraction for the transmitted light is

$$\theta_2 = 90.0° - \theta_p = 90.0° - 48.0° = 42.0° \quad\quad\quad ◊$$

---

59. Three polarizing plates whose planes are parallel are centered on a common axis. The directions of the transmission axes relative to the common vertical direction are shown in Figure P24.59. A linearly polarized beam of light with plane of polarization parallel to the vertical reference direction is incident from the left onto the first disk with intensity $I_i = 10.0$ units (arbitrary). Calculate the transmitted intensity I_f when $\theta_1 = 20.0°$, $\theta_2 = 40.0°$, and $\theta_3 = 60.0°$. *Hint:* Make repeated use of Malus's law.

Solution

When linearly polarized light of incident intensity I_0 passes through an analyzer whose transmission axis is rotated at angle θ to the plane of polarization of the incident light, the emerging light is linearly polarized in a direction parallel to the transmission axis of the analyzer. The intensity of the emerging light is given by

$$I = I_0 \cos^2 \theta \quad\quad \text{[Malus's law]}$$

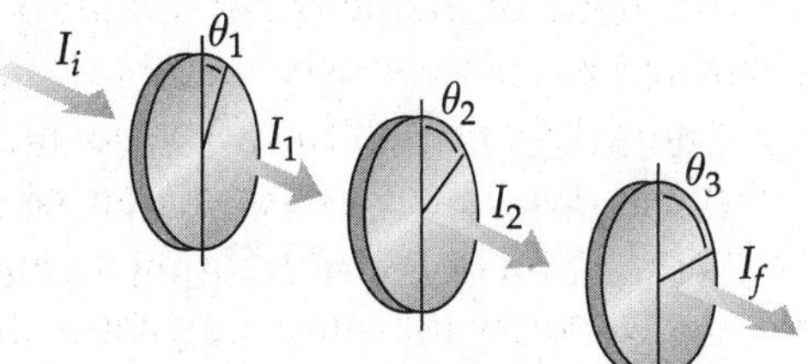

Figure P24.59

The first analyzer in Figure P24.59 has light of intensity I_i, polarized in the vertical plane, incident on it. The emerging light has intensity

$$I_1 = I_i \cos^2 \theta_1$$

The transmission axis of the second analyzer in the figure is at angle $\theta = \theta_2 - \theta_1$ from the plane of polarization of the light incident on it. The intensity of the light emerging from this analyzer is

$$I_2 = I_1 \cos^2(\theta_2 - \theta_1) = I_i \cos^2\theta_1 \cos^2(\theta_2 - \theta_1)$$

The third analyzer has its transmission axis at angle $\theta = \theta_3 - \theta_2$ from the plane of polarization of the light incident on it, so the emerging light has intensity

$$I_f = I_2 \cos^2(\theta_3 - \theta_2) = I_i \cos^2\theta_1 \cos^2(\theta_2 - \theta_1) \cos^2(\theta_3 - \theta_2)$$

If $I_i = 10.0$ units, $\theta_1 = 20.0°$, $\theta_2 = 40.0°$, and $\theta_3 = 60.0°$, the final emergent intensity is

$$I_f = (10.0 \text{ units}) \cos^2(20.0°) \cos^2(20.0°) \cos^2(20.0°) = 6.89 \text{ units} \qquad \lozenge$$

---

64. In a Young's interference experiment, the two slits are separated by 0.150 mm and the incident light includes two wavelengths: $\lambda_1 = 540$ nm (green) and $\lambda_2 = 450$ nm (blue). The overlapping interference patterns are observed on a screen 1.40 m from the slits. (a) Find a relationship between the orders m_1 and m_2 that determines where a bright fringe of the green light coincides with a bright fringe of the blue light. (The order m_1 is associated with λ_1 and m_2 with λ_2.) (b) Find the minimum values of m_1 and m_2 such that the overlapping of the bright fringes will occur, and the position of the overlap on the screen.

Solution

(a) In a double-slit interference pattern, the distance from the position of the central maximum to the location of the bright fringe of order m for light of wavelength λ is given by $y_m = m(\lambda L/d)$. Here, L is the distance from the plane of the slits to the plane of the viewing screen and d is the distance separating the two slits.

Therefore, if the bright line of order m_1 for wavelength λ_1 is to coincide with the bright line of order m_2 for wavelength λ_2, it is necessary that

$$m_1 \frac{\lambda_1 L}{d} = m_2 \frac{\lambda_2 L}{d} \qquad \text{or} \qquad m_1 \lambda_1 = m_2 \lambda_2$$

where both m_1 and m_2 are integers. With $\lambda_1 = 540$ nm and $\lambda_2 = 450$ nm, this becomes $(540 \text{ nm})m_1 = (450 \text{ nm})m_2$. Dividing both sides of this equation by 90 nm, we obtain the relation of the overlapping order numbers as

$$6m_1 = 5m_2 \qquad \lozenge$$

(b) Write the relation of part (a) as $m_2 = (6/5)m_1$ and look for integer values of m_1 that will yield an integer result for m_2. It is quickly discovered that the smallest such integer value for m_1 is $m_1 = 5$, which yields the integer result $m_2 = 6$. $\qquad \lozenge$

With $L = 1.40$ m and $d = 0.150$ mm, the distance on the screen between the central maximum and the position of the first overlap of bright fringes for the two given wavelengths is $y = m_1 \lambda_1 L/d = m_2 \lambda_2 L/d$, or

$$y = \frac{5(540 \times 10^{-9} \text{ m})(1.40 \text{ m})}{0.150 \times 10^{-3} \text{ m}} = \frac{6(450 \times 10^{-9} \text{ m})(1.40 \text{ m})}{0.150 \times 10^{-3} \text{ m}}$$

$$= 2.52 \times 10^{-2} \text{ m} = 2.52 \text{ cm} \qquad \lozenge$$

69. Figure P24.69 shows a radio-wave transmitter and a receiver, both $h = 50.0$ m above the ground and $d = 600$ m apart. The receiver can receive signals directly from the transmitter, and indirectly, from signals that bounce off the ground. If the ground is level between the transmitter and receiver, and a $\lambda/2$ phase shift occurs upon reflection, determine the longest wavelengths that interfere (a) constructively and (b) destructively.

Solution

The difference in the distances traveled by the direct and the indirect signals from the transmitter to the receiver is

$$\Delta d = 2d_1 - d$$

where

$$d_1 = \sqrt{(d/2)^2 + h^2} = \sqrt{d^2 + 4h^2}/2$$

or $\quad \Delta d = \sqrt{d^2 + 4h^2} - d$

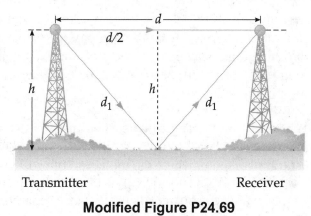

Transmitter   Receiver

Modified Figure P24.69

In addition, the reflected signal experiences a 180° phase shift (equivalent to a path difference of $\lambda/2$) upon reflection. Thus, the total difference in path lengths of the two signals is

$$\delta = \Delta d + \lambda/2 = \sqrt{d^2 + 4h^2} - d + \lambda/2$$

(a) For constructive interference, we must have $\delta = m\lambda$, where m is any integer. This gives

$$\sqrt{d^2 + 4h^2} - d = \left(m - \frac{1}{2}\right)\lambda \quad \text{or} \quad \lambda = \frac{2\left[\sqrt{d^2 + 4h^2} - d\right]}{2m - 1}$$

The longest wavelength satisfying this condition occurs for $m = 1$. With $h = 50.0$ m and $d = 600$ m, we find

$$\lambda_{\text{max}} = 2\left[\sqrt{(600 \text{ m})^2 + 4(50.0 \text{ m})^2} - 600 \text{ m}\right] = 16.6 \text{ m} \qquad \lozenge$$

(b) For destructive interference, it is necessary that $\delta = \left(m + 1/2\right)\lambda$, where m is a positive integer. Thus, $\sqrt{d^2 + 4h^2} - d = m\lambda$. The longest finite wavelength that can produce destructive interference corresponds to $m = 1$, which gives

$$\lambda_{max} = \sqrt{d^2 + 4h^2} - d = \sqrt{\left(600 \text{ m}\right)^2 + 4\left(50.0 \text{ m}\right)^2} - 600 \text{ m} = 8.28 \text{ m} \qquad \Diamond$$

---

73. A diffraction pattern is produced on a screen 140 cm from a single slit, using monochromatic light of wavelength 500 nm. The distance from the center of the central maximum to the first-order maximum is 3.00 mm. Calculate the slit width. *Hint:* Assume that the first-order maximum is halfway between the first- and second-order minima.

Solution

Dark fringes or minima in a single slit diffraction pattern exist at angles θ for which

$$\sin\theta = m\lambda/a \quad \text{with} \quad m = \pm 1, \pm 2, \pm 3, \dots$$

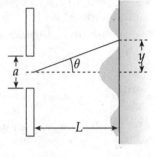

Here, θ is measured from the line passing through the center of the slit and perpendicular to the plane of the slit as shown in the figure, a is the width of the slit, and λ is the wavelength of the light illuminating the slit.

When $a \gg \lambda$, the angles where the minima occur are quite small, and we may use the approximation $\sin\theta \approx \tan\theta = y/L$. The screen position, measured from the location of the central maximum, where the minimum of order m occurs is then seen to be

$$y_m = L\tan\theta \approx L\sin\theta = L\left(\frac{m\lambda}{a}\right) \quad \text{or} \quad y_m = m\left(\frac{\lambda L}{a}\right)$$

Assuming the first-order maximum is halfway between the first- and second-order minima, its distance from the central maximum is

$$\left(y_{max}\right)_1 = y_1 + \frac{1}{2}\left(y_2 - y_1\right) = \frac{y_2 + y_1}{2} = \frac{1}{2}\left[2\left(\frac{\lambda L}{a}\right) + \left(\frac{\lambda L}{a}\right)\right] = \frac{3}{2}\left(\frac{\lambda L}{a}\right)$$

If it is observed that $\left(y_{max}\right)_1 = 3.00$ mm when $\lambda = 500$ nm and $L = 140$ cm, the width of the slit must be

$$a = \frac{3\lambda L}{2\left(y_{max}\right)_1} = \frac{3\left(500 \times 10^{-9} \text{ m}\right)\left(1.40 \text{ m}\right)}{2\left(3.00 \times 10^{-3} \text{ m}\right)} = 3.50 \times 10^{-4} \text{ m} = 0.350 \text{ mm} \qquad \Diamond$$

25

Optical Instruments

NOTES FROM SELECTED CHAPTER SECTIONS

25.2 The Eye

The following terms are used in describing the image-forming mechanism of the eye:

- **Cornea:** transparent structure through which light enters the eye.

- **Aqueous humor:** clear liquid behind the cornea.

- **Pupil:** an opening in the iris, dilates and contracts to regulate light.

- **Iris:** colored portion of the eye, muscular diaphragm that controls pupil size.

- **Retina, rods and cones:** back surface of eye, where images are formed.

- **Accommodation:** changing the focal length of the eye by action of the ciliary muscle.

- **Near point:** the closest distance for which the lens can accommodate to focus a sharp image on the retina (average value of around 25 cm).

- **Far point:** the greatest distance for which the lens in a relaxed eye can focus a clear image on the retina (in a normal eye the far point is at infinity).

- **Hyperopia** (farsightedness): condition caused either when the eyeball is too short or when the ciliary muscle is unable to change the shape of the lens enough to form a properly focused image of a nearby object.

- **Myopia** (nearsightedness): condition caused either when the eye is longer than normal or when the maximum focal length of the lens is insufficient to produce a clearly focused image of a distant object.

- **Presbyopia:** a reduction in accommodation ability resulting in the symptoms of farsightedness.

- **Astigmatism:** light from a point source produces a line image on the retina.

- **Diopters:** a measure of the power of a lens which equals the inverse of the focal length in meters.

25.4   The Compound Microscope

The overall magnification of a compound microscope of length L is equal to the product of the lateral magnification produced by the objective lens of focal length f_o (< 1 cm) and the angular magnification produced by the eyepiece lens of focal length f_e (a few centimeters). The two lenses, separated by a distance greater than either f_o or f_e, form an inverted, virtual image of an object.

25.5   The Telescope

There are two fundamentally different types of telescopes, both designed to aid in viewing distant objects, such as the planets in our solar system.

- The **refracting telescope** uses a combination of objective and eyepiece lenses to form an enlarged image. The two lenses are separated by a distance $L = f_e + f_o$.

- The **reflecting telescope** uses a concave mirror and an eyepiece lens to form an enlarged image.

For each type telescope, the angular magnification of the telescope equals the ratio of the focal length of the objective to the focal length of the eyepiece.

25.6   Resolution of Single-Slit and Circular Apertures

The wave nature of light limits, by diffraction, the minimum distance between objects which can be distinguished by an optical system.

Rayleigh's criterion determines the limiting condition for resolution of adjacent sources. Under this criterion two sources are just resolved when they are spaced so that the central maximum of the diffraction pattern of one source is located at the position of the first minimum of the diffraction pattern of the second source. Under this condition they are separated by the minimum spacing for which they can be seen as separate sources.

Objects S_1 and S_2 near minimum resolution.

The smallest angular separation that can be resolved is proportional to the wavelength of light used by the instrument and inversely proportional to the slit width or the diameter of a circular aperture.

The resolving power of a diffraction grating increases with the order number in the spectrum and is proportional to the number of lines illuminated in the grating. Refer to Equations (25.11) and (25.12) and consider the case of a diffraction grating in which 5 000 lines are illuminated.

In the zeroth order, $m = 0$ and all wavelengths are indistinguishable.

In the second order, $m = 2$ and the grating will produce a spectrum in which wavelengths differing in value by 1 part in 10 000 can be resolved.

25.7   The Michelson Interferometer

The Michelson interferometer can be used to measure distances to within a fraction of a wavelength of light. This is accomplished by splitting a light beam into two parts and then recombining the beams to form an interference pattern.

EQUATIONS AND CONCEPTS

The image forming properties of several optical instruments are described in the following paragraphs and equations.

Camera

The *f*-**number** of a lens is the ratio of its focal length to its diameter. This value is a measure of the light concentrating power of the lens and determines what is called the speed of the lens. *A fast lens has a small f-number, usually with a short focal length and a large diameter.*

$$f\text{-number} \equiv \frac{f}{D} \qquad (25.1)$$

Eye

The **power of a lens is measured in diopters** and equals the inverse of the focal length measured in meters. The correct algebraic sign must be used with *f*. *Optometrists and ophthalmologists usually prescribe lenses measured in diopters.*

$$\mathcal{P} = \frac{1}{f}$$

f is + for a converging lens and − for a diverging lens.

Simple magnifier

The **angular magnification of a simple magnifier** is the ratio of the angle subtended by an object when a lens is in use (θ) to the angle subtended by the object when it is placed at the near point of the eye (25 cm) with no lens (θ_0).

$$m \equiv \frac{\theta}{\theta_0} \qquad (25.2)$$

When the **image is at the near point** of the eye (25 cm), the angular magnification of a simple magnifier is maximum.

$$m_{\max} = 1 + \frac{25 \text{ cm}}{f} \qquad (25.5)$$

When the **image is at infinity** (most relaxed for the eye) the angular magnification is minimum. This is the expression for the magnification of a simple lens when the object is placed at the focal point of the lens.

$$m = \frac{25 \text{ cm}}{f} \qquad (25.6)$$

Compound microscope

The **overall magnification** is the product of the lateral magnification of the objective lens (M_1) and the angular magnification of the eyepiece (m_e). When an object is located just beyond the focal point of the objective, the two lenses in combination form an enlarged, virtual, and inverted image.

I_1 = image of object formed by objective lens.

I_2 = image of I_1 formed by the eyepiece.

$$M = M_1 m_e = -\frac{L}{f_o}\left(\frac{25\ cm}{f_e}\right) \qquad (25.7)$$

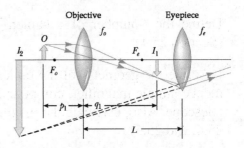

Astronomical telescope

The **angular magnification of a telescope** is equal to the ratio of the focal length of the objective to the focal length of the eyepiece. *The two converging lenses are separated by a distance equal to the sum of their focal lengths.*

$$m = \frac{f_0}{f_e} \qquad (25.8)$$

Rayleigh's criterion states the condition for the resolution of the images of two closely spaced sources.

For a **slit** the angular separation between sources (in radians) must be greater than the ratio of the wavelength of the light to the slit width, a.

$$\theta_{min} \cong \frac{\lambda}{a} \qquad (25.9)$$

For a **circular aperture,** the minimum angular separation which can be resolved depends on the wavelength of the light and the diameter of the aperture, D.

$$\theta_{min} = 1.22\frac{\lambda}{D} \qquad (25.10)$$

The **resolving power of a diffraction grating,** defined by Equation (25.11), increases as the number of lines illuminated, N, increases. Also, the resolving power is proportional to the order number, m, in which the spectrum is observed as stated in Equation (25.12).

$$R \equiv \frac{\lambda}{\lambda_2 - \lambda_1} = \frac{\lambda}{\Delta\lambda} \qquad (25.11)$$

where $\lambda \approx \lambda_1 \approx \lambda_2$

$$R = Nm \qquad (25.12)$$

REVIEW CHECKLIST

- Identify terms used in describing the image-forming mechanism of the eye. Describe the use of lenses in the correction of common defects of the eye. (Section 25.2)

- Define the *f*-number of a camera lens and relate this criterion to shutter speed. (Section 25.1)

- Describe the geometry of the optical components for each of several optical instruments: simple magnifier, compound microscope, reflecting telescope, and refracting telescope. Also, calculate the magnifying power for each instrument. (Sections 25.3, 25.4, and 25.5)

- Determine whether or not two sources under a given set of conditions are resolvable as defined by Rayleigh's criterion. (Section 25.6)

- Understand what is meant by the resolving power of a grating, and calculate the resolving power of a grating under specified conditions. (Section 25.6)

- Describe the technique employed in the Michelson interferometer for precise measurement of length based on known values for the wavelength of light. (Section 25.7)

SOLUTIONS TO SELECTED END-OF-CHAPTER PROBLEMS

3. A photographic image of a building is 0.092 0 m high. The image was made with a lens with a focal length of 52.0 mm. If the lens was 100 m from the building when the photograph was made, determine the height of the building.

Solution

From the thin lens equation, we find

$$q = \frac{pf}{p-f} = \frac{(100 \text{ m})(52.0 \times 10^{-3} \text{ m})}{100 \text{ m} - 52.0 \times 10^{-3} \text{ m}} = 5.203 \times 10^{-2} \text{ m} \approx f$$

Consider the figure below, showing two rays passing undeviated through the center of the lens to the top and bottom of the image.

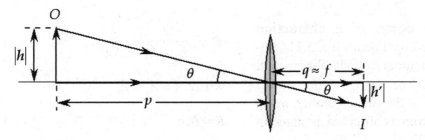

Using the properties of similar triangles, we write $|h|/|h'| = p/f$, or the size of the object must be

$$|h| = |h'|\left(\frac{p}{f}\right) = (0.092\ 0\ \text{m})\left(\frac{100\ \text{m}}{52.0 \times 10^{-3}\ \text{m}}\right) = 177\ \text{m} \qquad \Diamond$$

Alternatively, one could have used the lateral magnification $M = h'/h = -q/p$ to obtain

$$h = \frac{h'}{-q/p} \approx -h'\left(\frac{p}{f}\right) \qquad \text{or} \qquad |h| = |h'|\left(\frac{p}{f}\right)$$

which is the same as what we had above, and which yields the same result. The negative sign in the magnification equation simply tells us that the image is inverted in this situation.

---

7. A certain type of film requires an exposure time of 0.010 s with an $f/11$ lens setting. Another type of film requires twice the light energy to produce the same level of exposure. What f-number does the second type of film need with the 0.010-s exposure time?

Solution

The energy delivered to the film by the light striking it is $\Delta E = \mathcal{P}(\Delta t) = IA(\Delta t)$, where I is the intensity of the light, A is the area of the film that is exposed, and Δt is the exposure time.

Since the exposure time and the area of film exposed (that is, the size of the image) is the same with the two films, we see that if we are to double the energy delivered to the film, the intensity of the light reaching the film must be doubled.

The intensity of light reaching the film is proportional to the reciprocal of the square of the f-stop setting. (See the statement of Problem 6 in Chapter 25 of the textbook). That is,

$$I = \frac{constant}{(f\text{-number})^2}$$

Thus, if the intensity I_2 of the light exposing the second film is to be double the intensity I_1 used to expose the first film, it is necessary that

$$\frac{constant}{(f\text{-number})_2^2} = 2\left[\frac{constant}{(f\text{-number})_1^2}\right]$$

or $\quad (f\text{-number})_2^2 = \dfrac{(f\text{-number})_1^2}{2} = \dfrac{(11)^2}{2} = 61$, giving $(f\text{-number})_2 = 7.8$

This means you should use the $f/8$ setting on your camera for the new film. $\qquad \Diamond$

---

11. The accommodation limits for Nearsighted Nick's eyes are 18.0 cm and 80.0 cm. When he wears his glasses, he is able to see faraway objects clearly. At what minimum distance is he able to see objects clearly?

Solution

In order for Nick to see distant objects clearly, the lens in his glasses must form an upright, virtual image of very distant objects $(p \rightarrow \infty)$ at the far point of his eyes (that is, the image distance must be $q = -80.0$ cm).

The thin lens equation then gives the required focal length for the lens in his glasses as

$$\frac{1}{f} = \frac{1}{p} + \frac{1}{q} = \frac{1}{\infty} + \frac{1}{-80.0 \text{ cm}}$$

or

$$f = -80.0 \text{ cm}$$

When Nick uses his glasses to observe nearby objects, the lens must form an erect, virtual image at the near point of his eye of the closest objects viewed clearly. That is, when $p = p_{min}$, it is necessary that $q = q_{min} = -18.0$ cm. The thin lens equation then gives the minimum viewing distance as

$$\frac{1}{p_{min}} = \frac{1}{f} - \frac{1}{q_{min}}$$

or

$$p_{min} = \frac{f\, q_{min}}{q_{min} - f} = \frac{(-80.0 \text{ cm})(-18.0 \text{ cm})}{-18.0 \text{ cm} - (-80.0 \text{ cm})} = 23.2 \text{ cm} \qquad \Diamond$$

14. A patient has a near point of 45.0 cm and far point of 85.0 cm. (a) Can a single lens correct the patient's vision? Explain the patient's options. (b) Calculate the power lens needed to correct the near point, so the patient can see objects 25.0 cm away. Neglect the eye-lens distance. (c) Calculate the power lens needed to correct the patient's far point, again neglecting the eye-lens distance.

Solution

(a) This patient cannot focus on objects that are farther than 85.0 cm from the eye, and needs corrective action for the distant vision. Also, the patient is unable to focus on objects that are less than 45.0 cm from the eye, and requires corrective action in the near vision. Both corrective actions can be accomplished using a single bifocal lens, with the upper portion having the focal length (or optical power) needed to correct the far vision, while the lower portion of the lens has the required focal length or power to correct the near vision. Alternately, the patient must purchase two pairs of glasses, one pair for reading, and a second pair for distant vision. $\qquad \Diamond$

(b) To correct the near vision, the patient should be able to clearly see objects located at the near point for a normal eye ($p = 25.0$ cm). The corrective lens must form an upright, virtual image located at the near point of the patient's eye ($q = -45.0$ cm) for such objects. The thin lens equation gives the required focal length for the lens as

$$f = \frac{pq}{p+q} = \frac{(25.0 \text{ cm})(-45.0 \text{ cm})}{25.0 \text{ cm} - 45.0 \text{ cm}} = +56.3 \text{ cm} = 0.563 \text{ m}$$

The power, in diopters, of this lens would be

$$\mathcal{P} = \frac{1}{f_{\text{in meters}}} = \frac{1}{+0.563 \text{ m}} = +1.78 \text{ diopters} \qquad\qquad \lozenge$$

(c) To correct the distant vision, the corrective lens must form upright, virtual images located at the eye's far point ($q = -85.0$ cm) for very distant objects ($p \rightarrow \infty$). The required focal length is given by the thin lens equation as

$$\frac{1}{f} = \frac{1}{p} + \frac{1}{q} \rightarrow \frac{1}{\infty} + \frac{1}{-85.0 \text{ cm}} \qquad \text{or} \qquad f = -85.0 \text{ cm} = -0.850 \text{ m}$$

and the needed power is

$$\mathcal{P} = \frac{1}{f_{\text{in meters}}} = \frac{1}{-0.850 \text{ m}} = -1.18 \text{ diopters} \qquad\qquad \lozenge$$

═══════════════════

23. A leaf of length h is positioned 71.0 cm in front of a converging lens with a focal length of 39.0 cm. An observer views the image of the leaf from a position 1.26 m behind the lens, as shown in Figure P25.23. (a) What is the magnitude of the lateral magnification (the ratio of the image size to the object size) produced by the lens? (b) What angular magnification is achieved by viewing the image of the leaf rather than viewing the leaf directly?

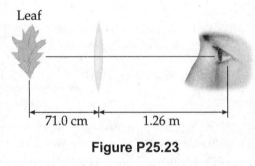

Figure P25.23

Solution

(a) With a focal length of $f = +39.0$ cm and an object distance of $p = +71.0$ cm, the thin lens equation gives the image distance as

$$q = \frac{pf}{p-f} = \frac{(71.0 \text{ cm})(39.0 \text{ cm})}{71.0 \text{ cm} - 39.0 \text{ cm}} = +86.5 \text{ cm}$$

Thus, the image of the leaf is 86.5 cm to the right of the lens or 39.5 cm in front of the observer's eye. The lateral magnification by the lens is

$$M = \frac{h'}{h} = -\frac{q}{p} = -\frac{86.5 \text{ cm}}{71.0 \text{ cm}} = -1.22 \qquad\qquad \lozenge$$

(b) If the unaided eye viewed the leaf, of height h, directly from a distance of $d = 126 \text{ cm} + 71.0 \text{ cm} = 197 \text{ cm}$, the angular size of the leaf would be

$$\theta_0 = \frac{h}{d} = \frac{h}{197 \text{ cm}}$$

If instead, the eye views the image formed by the lens (located 86.5 cm beyond the lens and of height $h' = |M|h = 1.22h$) from a distance of $d' = 126 \text{ cm} - 86.5 \text{ cm} = 39.5 \text{ cm}$, the angular size of the object the eye is viewing would be

$$\theta = \frac{h'}{d'} = \frac{1.22h}{39.5 \text{ cm}}$$

so the angular magnification achieved by viewing the image rather than the leaf directly is

$$m = \frac{\theta}{\theta_0} = \frac{1.22h/39.6 \text{ cm}}{h/197 \text{ cm}} = 1.22 \left(\frac{197 \text{ cm}}{39.5 \text{ cm}} \right) = 6.08 \qquad \lozenge$$

---

28. A microscope has an objective lens with a focal length of 16.22 mm and an eyepiece with a focal length of 9.50 mm. With the length of the barrel set at 29.0 cm, the diameter of a red blood cell's image subtends an angle of **1.43 mrad** with the eye. If the final image distance is 29.0 cm from the eyepiece, what is the actual diameter of the red blood cell?

Solution

We are told that the final image of the blood cell is located 29.0 cm in front of the eye and that it intercepts an angle of **1.43 mrad**. The diameter of this final image must be

$$h_e = r\theta = \left(29.0 \times 10^{-2} \text{ m}\right)\left(1.43 \times 10^{-3} \text{ rad}\right) = 4.15 \times 10^{-4} \text{ m}$$

At this point, one is tempted to use Equation 25.7 from the textbook for the overall magnification of a compound microscope, and compute $h = h_e/m$ as the size of the blood cell serving as the object for the microscope. However, the derivation of that equation is based on several assumptions, one of which is that the eye is relaxed and viewing a final image located an infinite distance in front of the eyepiece. This is clearly not true in this case, and the use of Equation (25.7) would introduce considerable error. Instead, we return to basics and use the thin lens equation to find the size of the original object.

The object for the eyepiece is the image formed by the objective lens, and we shall call the size of this image h'. The lateral magnification produced by the eyepiece is $M_e = h_e/h' = -q_e/p_e$, where $q_e = -29.0 \text{ cm}$ and p_e is the object distance for the eyepiece. The magnification produced by the objective lens is $M_1 = h'/h = -q_1/p_1$, with $q_1 = L - p_e$ (where $L = 29.0 \text{ cm}$ is the distance between the lenses of the microscope) and p_1 is the object distance for the objective lens. The overall magnification is

$$M = \frac{h_e}{h} = \left(\frac{h'}{h}\right)\left(\frac{h_e}{h'}\right) = M_1 M_e$$

and the actual size of the blood cell is $h = h_e/|M| = (4.15 \times 10^{-4} \text{ m})/|M|$.

From the thin lens equation, the object distance for the eyepiece must be

$$p_e = \frac{q_e f_e}{q_e - f_e} = \frac{(-29.0 \text{ cm})(0.950 \text{ cm})}{-29.0 \text{ cm} - 0.950 \text{ cm}} = +0.920 \text{ cm}$$

and

$$M_e = -\frac{q_e}{p_e} = -\frac{-29.0 \text{ cm}}{0.920 \text{ cm}} = +31.5$$

The image distance for the objective lens is then $q_1 = 29.0 \text{ cm} - 0.920 \text{ cm} = 28.1 \text{ cm}$, and the thin lens equation gives the object distance for this lens as

$$p_1 = \frac{q_1 f_o}{q_1 - f_o} = \frac{(28.1 \text{ cm})(1.622 \text{ cm})}{28.1 \text{ cm} - 1.622 \text{ cm}} = +1.72 \text{ cm}$$

Thus, $\quad M_1 = -\frac{q_1}{p_1} = -\frac{28.1 \text{ cm}}{1.72 \text{ cm}} = -16.3 \quad$ and $\quad M = M_1 M_e = (-16.3)(31.5) = -513$

The actual diameter of the blood cell must be

$$h = \frac{h_e}{|M|} = \frac{4.15 \times 10^{-4} \text{ m}}{513} = 8.09 \times 10^{-7} \text{ m} = 0.809 \ \mu\text{m} \qquad \lozenge$$

---

35. A person decides to use an old pair of eyeglasses to make some optical instruments. He knows that the near point in his left eye is 50.0 cm and the near point in his right eye is 100 cm. (a) What is the maximum angular magnification he can produce in a telescope? (b) If he places the lenses 10.0 cm apart, what is the maximum overall magnification he can produce in a microscope? (To solve part (b), go back to basics and use the thin lens equation.)

Solution

To allow a farsighted person to see objects at a normal reading distance clearly, the eyeglass lenses form upright, virtual images at the near point of each eye when the object is 25.0 cm in front of the lens. Thus, we use the thin lens equation to find the focal length of each lens. For the left-side lens, using $q_L = -50.0 \text{ cm}$ when $p_L = 25.0 \text{ cm}$ gives

$$f_L = \frac{p_L q_L}{p_L + q_L} = \frac{(25.0 \text{ cm})(-50.0 \text{ cm})}{25.0 \text{ cm} - 50.0 \text{ cm}} = 50.0 \text{ cm}$$

Similarly, for the right-side lens, using $q_R = -100 \text{ cm}$ when $p_R = 25.0 \text{ cm}$ gives $f_R = 33.3 \text{ cm}$.

(a) The angular magnification produced by an astronomical telescope is $m = f_{\text{objective}} / f_{\text{eyepiece}}$. To achieve maximum magnification, one should use the longer focal length lens as the objective and the shorter focal length lens as the eyepiece of the telescope. This yields a magnification of

$$m = \frac{f_L}{f_R} = \frac{50.0 \text{ cm}}{33.3 \text{ cm}} = 1.50 \qquad \lozenge$$

(b) Using the right-side lens as the eyepiece and, for maximum magnification, requiring the final image be formed as close as a normal eye can focus $(q_e = -25.0 \text{ cm})$ gives the angular magnification by the eyepiece as

$$m_{e,\text{max}} = 1 + \frac{25.0 \text{ cm}}{f_e} = 1 + \frac{25.0 \text{ cm}}{33.3 \text{ cm}} = 1.75$$

and the object distance for the eyepiece as

$$p_e = \frac{q_e f_e}{q_e - f_e} = \frac{(-25.0 \text{ cm})(33.3 \text{ cm})}{-25.0 \text{ cm} - 33.3 \text{ cm}} = 14.3 \text{ cm}$$

Since the object for the eyepiece is the image formed by the objective lens, the image distance for the objective lens must be

$$q_1 = L - p_e = 10.0 \text{ cm} - 14.3 \text{ cm} = -4.3 \text{ cm}$$

Therefore, the objective lens (the left-side lens from the eyeglasses) must form a virtual image 4.3 cm in front of it. This requires an object distance of

$$p_1 = \frac{q_1 f_o}{q_1 - f_o} = \frac{(-4.3 \text{ cm})(50.0 \text{ cm})}{-4.3 \text{ cm} - 50.0 \text{ cm}} = 4.0 \text{ cm}$$

The maximum overall magnification from a 10.0 cm long compound microscope made from the available lenses is then

$$m_{\text{max}} = M_1 m_{e,\text{max}} = \left(-\frac{q_1}{p_1}\right) m_{e,\text{max}} = \left(-\frac{-4.3 \text{ cm}}{4.0 \text{ cm}}\right)(1.75) = 1.9 \qquad \lozenge$$

Repeating the calculation of part (b) using the left-side lens for the eyepiece yields an overall magnification of $m = 1.8$, so our original choice for the eyepiece lens was the better choice.

---

41. A vehicle with headlights separated by 2.00 m approaches an observer holding an infrared detector sensitive to radiation of wavelength 885 nm. What aperture diameter is required in the detector if the two headlights are to be resolved at a distance of 10.0 km?

Solution

Any time light from a source passes through an aperture, diffraction effects occur. If the aperture is circular, the resulting diffraction pattern is a bright central spot surrounded by concentric alternating bright and dark rings. When light from two sources having a small angular separation passes through the aperture, the diffraction patterns overlap, often making it difficult to recognize that the light originated from two distinct sources.

If two headlights, sources S_1 and S_2, have a lateral separation of $\Delta s = 2.00$ m and are viewed from a distance of $r = 10.0$ km, their angular separation is

$$\theta = \frac{\Delta s}{r} = \frac{2.00 \text{ m}}{10.0 \times 10^3 \text{ m}} = 2.00 \times 10^{-4} \text{ radians}$$

After the light passes through the aperture of a detector, the images are considered just resolved, according to Rayleigh's criterion, if the central maximum in the diffraction pattern of one image falls on the first minimum of the diffraction pattern of the other image as shown in the sketch at the right.

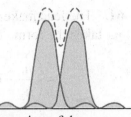

To just meet Rayleigh's criterion with a circular aperture, the angular separation of the sources must equal a minimum value given by $\theta_{min} = 1.22\lambda/D$ where λ is the wavelength of the light and D is the diameter of the aperture. Thus, for the images of the headlights to be just resolved by the infrared detector (with $\theta = \theta_{min}$), the required diameter of the detector aperture is

$$D = \frac{1.22\lambda}{\theta} = \frac{1.22\left(885 \times 10^{-9} \text{ m}\right)}{2.00 \times 10^{-4} \text{ rad}} = 5.40 \times 10^{-3} \text{ m} = 5.40 \text{ mm}$$ ◊

45. A 15.0-cm-long grating has 6 000 slits per centimeter. Can two lines of wavelengths 600.000 nm and 600.003 nm be separated with this grating? Explain.

Solution

The resolving power required to separate the two specified wavelengths is

$$R_{required} = \frac{\lambda}{\Delta\lambda} = \frac{600.000 \text{ nm}}{\left(600.003 - 600.000\right) \text{ nm}} = 2.00 \times 10^{5}$$

The resolving power of a diffraction grating in the mth order is $R = Nm$, where N is the total number of slits on the grating. The number of slits on the specified grating is

$$N = \left(15.0 \text{ cm}\right)\left(6\,000 \text{ slits/cm}\right) = 9.00 \times 10^{4}$$

The spacing between adjacent slits on this grating is

$$d = \frac{1.00 \text{ cm}}{6\,000} = 1.67 \times 10^{-4} \text{ cm} = 1.67 \times 10^{-6} \text{ m}$$

Because the angle of diffraction cannot exceed 90°, the grating equation gives the maximum order of 600 nm light one could hope to observe with this grating as

$$m = \frac{d \sin\theta_{max}}{\lambda} = \frac{\left(1.67 \times 10^{-6} \text{ m}\right)\sin 90°}{600 \times 10^{-9} \text{ m}} = 2.78$$

The order number must have integer values, and the third order cannot be reached. Therefore, when using this grating, the last observable order in the 600 nm region of the spectrum is the second order $\left(m_{max} = 2\right)$. The maximum resolving power in this region for the specified grating is then

$$R_{max} = Nm_{max} = \left(9.00 \times 10^{4}\right)\left(2\right) = 1.80 \times 10^{5}$$

Since this value is less than the required resolving power, it is not possible to separate the two wavelengths in question using this grating.   ◊

52. The Michelson interferometer can be used to measure the index of refraction of a gas by placing an evacuated transparent tube in the light path along one arm of the device. Fringe shifts occur as the gas is slowly added to the tube. Assume that 600-nm light is used, that the tube is 5.00 cm long, and that 160 fringe shifts occur as the pressure of the gas in the tube increases to atmospheric pressure. What is the index of refraction of the gas? *Hint:* The fringe shifts occur because the wavelength of the light changes inside the gas-filled tube.

Solution

In the fringe pattern produced by a Michelson interferometer, a fringe shift occurs each time the effective length of one arm of the interferometer changes by a quarter-wavelength. In this case, the effective length of the arm changes because the number of wavelengths that fit between the ends of the tube changes as the tube fills with gas. Four fringe shifts occur for each additional wavelength that can be fitted within the length of the tube.

Thus, if L is the length of the tube, λ is the wavelength of the light in a vacuum, and $\lambda_n = \lambda/n_{gas}$ is the wavelength of the light when traveling in a gas having refractive index n_{gas}, the number of shifts one should observe as the tube is filling with gas is

$$N = 4\left[\frac{L}{\lambda_n} - \frac{L}{\lambda}\right] = 4\left[\frac{L}{\lambda/n_{gas}} - \frac{L}{\lambda}\right] = \frac{4L}{\lambda}\left(n_{gas} - 1\right)$$

The index of refraction of the gas used to fill the tube must then be

$$n_{gas} = 1 + \frac{N\lambda}{4L}$$

For the given situation, $L = 5.00$ cm and $\lambda = 600$ nm. Thus, if 160 fringe shifts are observed as the tube fills, the index of refraction of the gas in the tube is

$$n_{gas} = 1 + \frac{160\left(600 \times 10^{-9} \text{ m}\right)}{4\left(5.00 \times 10^{-2} \text{ m}\right)} = 1.000\ 5$$ ◊

---

57. The near point of an eye is 75.0 cm. (a) What should be the power of a corrective lens prescribed to enable the eye to see an object clearly at 25.0 cm? (b) If, using the corrective lens, the person can see an object clearly at 26.0 cm, but not at 25.0 cm, by how many diopters did the lens grinder miss the prescription?

Solution

(a) If the corrective lens is to work properly, it should form an upright, virtual image at the near point of the eye ($q = -75.0$ cm) when the object is located 25.0 cm in front of the eye ($p = +25.0$ cm).

The thin lens equation gives the required optical power of the corrective lens as

$$\mathcal{P}_{required} = \underbrace{\frac{1}{f_{in}}}_{\text{meters}} = \frac{1}{p} + \frac{1}{q} = \frac{1}{0.250 \text{ m}} - \frac{1}{0.750 \text{ m}} = \frac{3-1}{0.750 \text{ m}}$$

or $\quad \mathcal{P}_{required} = \dfrac{2}{0.750 \text{ m}} = +2.67$ diopters ◊

(b) If the user actually sees a clear image (located at the near point with $q = -75.0$ cm) when the object is 26.0 cm in front of the eye rather than the desired 25.0 cm, the actual power of the provided corrective lens is

$$\mathcal{P}_{actual} = \frac{1}{f} = \frac{1}{p} + \frac{1}{q} = \frac{1}{0.260 \text{ m}} - \frac{1}{0.750 \text{ m}} = \frac{0.750 \text{ m} - 0.260 \text{ m}}{(0.260 \text{ m})(0.750 \text{ m})}$$

or $$\mathcal{P}_{actual} = \frac{+0.490 \text{ m}}{(0.260 \text{ m})(0.750 \text{ m})} = +2.51 \text{ diopters}$$

The magnitude of the error made in grinding this corrective lens is then seen to be

$$\Delta \mathcal{P} = \mathcal{P}_{required} - \mathcal{P}_{actual} = 2.67 \text{ diopters} - 2.51 \text{ diopters} = 0.16 \text{ diopters}$$

or the supplied lens has a power that is 0.16 diopters too low.   ◊

---

61. A boy scout starts a fire by using a lens from his eyeglasses to focus sunlight on kindling 5.0 cm from the lens. The boy scout has a near point of 15 cm. When the lens is used as a simple magnifier, (a) what is the maximum magnification that can be achieved, and (b) what is the magnification when the eye is relaxed? *Caution:* The equations derived in the text for a simple magnifier assume a "normal" eye.

Solution

The scout started the fire by forming an image of the Sun (a very distant object for which $p \to \infty$) on the kindling with an image distance of $q = 5.0$ cm. The thin lens equation then gives the focal length of the lens as

$$\frac{1}{f} = \frac{1}{p} + \frac{1}{q} \approx 0 + \frac{1}{q} \qquad \text{or} \qquad f \approx q = +5.0 \text{ cm}$$

When the object is viewed most clearly with the unaided eye, it is located at the near point (at a distance of $d = 15$ cm for this eye) and has an angular size of

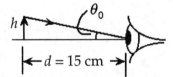

$$\theta_0 \approx \tan \theta_0 = \frac{h}{d} \qquad \text{or} \qquad \theta_0 = \frac{h}{15 \text{ cm}}$$

(a) For maximum magnification, the magnifier must form a magnified, upright, virtual image at the near point of the eye ($q = -15$ cm for this eye). To achieve this, the thin lens equation gives the reciprocal of the required object distance as

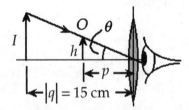

$$\frac{1}{p} = \frac{1}{f} - \frac{1}{q} = \frac{1}{f} + \frac{1}{15 \text{ cm}}$$

and the angular size of the magnified image viewed by the eye is

$$\theta \approx \tan\theta = \frac{h}{p} = h\left(\frac{1}{p}\right) = h\left(\frac{1}{f} + \frac{1}{15 \text{ cm}}\right)$$

The maximum magnification is then

$$m_{\text{max}} = \frac{\theta}{\theta_0} = \frac{h(1/f + 1/15 \text{ cm})}{h/15 \text{ cm}} = \not{h}\left(\frac{1}{f} + \frac{1}{15 \text{ cm}}\right)\frac{15 \text{ cm}}{\not{h}} = \frac{15 \text{ cm}}{f} + 1$$

or $$m_{\text{max}} = \frac{15 \text{ cm}}{5.0 \text{ cm}} + 1 = 4.0$$ ◊

(b) When the magnifier is positioned for most comfortable viewing with a relaxed eye, the object is at the focal point ($p = f$) and the virtual image recedes to infinity with parallel rays entering the eye.

Consider the second drawing given above and recognize that, when $p = f$, the angular size of the image viewed by the eye is $\theta \approx \tan\theta = h/f$ and the magnification is

$$m = \frac{\theta}{\theta_0} = \frac{h/f}{h/15 \text{ cm}} = \frac{15 \text{ cm}}{f} = \frac{15 \text{ cm}}{5.0 \text{ cm}} = 3.0$$ ◊

26

Relativity

NOTES FROM SELECTED CHAPTER SECTIONS

26.1 Galilean Relativity

According to the principle of Galilean relativity, the laws of mechanics are the same in all inertial frames of reference. *Inertial frames of reference are those coordinate systems which are at rest with respect to one another or which move at constant velocity with respect to one another. There is no preferred frame of reference for describing the laws of mechanics.*

26.2 The Speed of Light

The Michelson-Morley experiment was designed to detect the velocity of the Earth with respect to the hypothetical luminiferous ether and thereby confirm the prediction of an absolute frame of reference. The instrument used is called an interferometer (refer to Chapter 25, Section 25.7), and the measurement involved an attempt to observe a fringe shift in the interference pattern of two light beams. The light beams were directed along perpendicular paths, one parallel to, and one perpendicular to the "ether wind" and then combined to form the interference pattern. *The outcome of the experiment was negative,* contradicting the ether hypothesis. *Light is now understood to be an electromagnetic wave that requires no medium for its propagation.*

26.3 Einstein's Principle of Relativity

Einstein's special theory of relativity is based upon two postulates:

1. The laws of physics are the same in all inertial frames of reference. *An inertial frame of reference is a non-accelerated frame.*

2. The speed of light in a vacuum has the same value as measured by all observers in all inertial reference frames. *The measured value for the speed of light is independent of the motion of the observer or of the motion of the light source.*

26.4 Consequences of Special Relativity

In relativistic mechanics length and time are not absolute. Following are some of the terms used to describe the consequences of special relativity:

- **Simultaneity:** Two events that are simultaneous in one reference frame are, in general, not simultaneous in another frame which is moving with respect to the first.

- **Proper time:** The time interval between two events as measured by an observer who sees the two events occur at the same position.

- **Time dilation:** A moving clock runs slower than an identical clock at rest with respect to the observer.

- **Proper length:** The length of an object measured by an observer at rest relative to the object (moving with the object).

- **Length contraction:** The length of an object measured in a reference frame that is moving with respect to the object is always less than the proper length if this length is oriented parallel to the motion. *Lengths oriented perpendicular to the motion do not exhibit length contraction.*

26.5 Relativistic Momentum

To account for relativistic effects, it is necessary to modify the classical definition of momentum to satisfy the following conditions:

- The relativistic momentum must be conserved in all inertial frames.

- The relativistic momentum must approach the classical value, *mv*, as the ratio *v/c* approaches zero.

26.6 Relativistic Energy and the Equivalence of Mass and Energy

A particle has a rest energy proportional to the value of its inertial mass ($E_R = mc^2$). When the particle is in motion, it has a total energy equal to the sum of its kinetic energy and its rest energy.

26.7 General Relativity

Einstein's postulates of general relativity are:

- All the laws of nature have the same form for observers in any reference frame (inertial or non-inertial).

- In the vicinity of any given point, a gravitational field is equivalent to an accelerated frame of reference without a gravitational field (the principle of equivalence).

EQUATIONS AND CONCEPTS

Basic postulates of the special theory of relativity are:

- The laws of physics are the same in all inertial frames of reference.

- The speed of light has the same value in all inertial frames.

Important consequences of the theory of special relativity:

- **Time dilation** — A time interval Δt measured by an observer moving with respect to a clock is longer than the time interval Δt_p (the proper time) measured by an observer at rest with respect to the clock. *Moving clocks run slower than clocks at rest with respect to an observer.*

$$\Delta t = \frac{\Delta t_p}{\sqrt{1 - v^2/c^2}} = \gamma \, \Delta t_p \qquad (26.2)$$

$$\gamma = \frac{1}{\sqrt{1 - v^2/c^2}} \qquad (26.3)$$

- **Length contraction** — If an object has a length L_p (the proper length) when measured by an observer at rest with respect to the object, the length L measured by an observer moving with respect to the object will be less than L_p. *Length contraction occurs only along the direction of motion.*

$$L = \frac{L_p}{\gamma} = L_p\sqrt{1 - \frac{v^2}{c^2}} \qquad (26.4)$$

- **Simultaneity** — events observed as simultaneous in one frame of reference are not necessarily observed as simultaneous in a second frame moving relative to the first.

Some important equations of relativistic mechanics:

- The **relativistic linear momentum** of a particle with mass m and moving with speed v satisfies the following conditions:

$$p \equiv \frac{mv}{\sqrt{1 - v^2/c^2}} = \gamma mv \qquad (26.5)$$

  (1) Momentum is conserved in all collisions.

  (2) The relativistic value of momentum approaches the classical value (mv) as v approaches zero.

- The **relativistic kinetic energy** of a particle of mass m moving with a speed v, includes the **rest energy** term mc^2.

$$KE = \gamma mc^2 - mc^2 \qquad (26.6)$$
$$KE = (\gamma - 1)mc^2$$

- The **rest energy** of a particle is independent of the speed of the particle. The mass m must have the same value in all inertial frames.

$$E_R = mc^2 \qquad (26.7)$$

- The **total energy** E of a particle is the sum of the kinetic energy and the rest energy. *This expression shows that mass is a form of energy.* Equation (26.10) is useful in cases where the speed of an object is not known.

$$E = KE + mc^2 = \gamma mc^2 \qquad (26.8)$$
$$E = \frac{mc^2}{\sqrt{1 - v^2/c^2}} \qquad (26.9)$$
$$E^2 = p^2c^2 + (mc^2)^2 \qquad (26.10)$$

Photons ($m = 0$) travel with the speed of light. Equation (26.11) is an exact expression relating energy and momentum for particles which have zero mass.

$$E = pc \qquad (26.11)$$

The **electron volt** (eV) is a convenient energy unit when stating the energies of electrons and other subatomic particles.

$$1 \text{ eV} = 1.60 \times 10^{-19} \text{ J}$$
$$m_e c^2 = 0.511 \text{ MeV}$$
$$= 8.20 \times 10^{-14} \text{ J}$$

REVIEW CHECKLIST

- Understand the Michelson-Morley experiment, its objectives, results, and the significance of its outcome. (Section 26.2)

- State Einstein's two postulates of the special theory of relativity. (Section 26.3)

- Understand the idea of simultaneity and the fact that simultaneity is not an absolute concept. That is, two events which are simultaneous in one reference frame are not simultaneous when viewed from a second frame moving with respect to the first. (Section 26.4)

- Make calculations using the equations for time dilation and length contraction. (Section 26.4)

- State the correct relativistic expressions for the momentum, kinetic energy, and total energy of a particle. Make calculations using these equations. (Sections 26.5 and 26.6)

SOLUTIONS TO SELECTED END-OF-CHAPTER PROBLEMS

1. If astronauts could travel at $v = 0.950 c$, we on Earth would say it takes $(4.20/0.950) = 4.42$ years to reach Alpha Centauri, 4.20 lightyears away. The astronauts disagree. (a) How much time passes on the astronauts' clocks? (b) What is the distance to Alpha Centauri as measured by the astronauts?

Solution

(a) Observers on Earth measure the distance to Alpha Centauri as $d = 4.20$ ly, where $1 \text{ ly} = c(1 \text{ year})$. These observers see the ship moving toward Alpha Centauri at a constant speed of $v = 0.950 c$, and compute the required travel time to be $\Delta t = d/v = 4.20 \text{ ly}/0.950 c = 4.42$ yr. But these observers are in motion relative to the astronaut's internal biological clocks and hence experience a dilated version of the proper time interval Δt_p measured on those clocks. From the time dilation equation, $\Delta t = \gamma \Delta t_p$ where $\gamma = 1 / \sqrt{1 - (v/c)^2}$, we find the time that seems to have passed on the astronaut's clock during the trip to be

$$\Delta t_p = \frac{\Delta t}{\gamma} = \Delta t \sqrt{1 - (v/c)^2} = (4.42 \text{ yr}) \sqrt{1 - (0.950)^2} = 1.38 \text{ yr} \qquad \lozenge$$

(b) Since any motion of Earth occurs at a speed very small in comparison to the speed of light, we consider observers on Earth at rest relative to the span of space separating Earth and Alpha Centauri. Then, the length they measure for the required travel distance is the proper length, or $L_p = d = 4.20$ ly. The astronauts are moving at $v = 0.950c$ relative to this span, and their measure of the travel distance is length contracted as given by

$$L = L_p/\gamma = L_p\sqrt{1-(v/c)^2} = (4.20 \text{ ly})\sqrt{1-(0.950)^2} = 1.31 \text{ ly}$$ ◊

Now, we can note that the astronauts looking out the front window of the spacecraft see Alpha Centauri approaching them at $v = 0.950c$, and they detect an elapsed time $t = L/v = 1.31 \text{ ly}/0.950c = 1.38$ yr as being required to complete the trip. This is consistent with the answer from part (a).

---

5. The average lifetime of a pi meson in its own frame of reference (i.e., the proper lifetime) is 2.6×10^{-8} s. If the meson moves with a speed of $0.98\ c$, what is (a) its mean lifetime as measured by an observer on Earth, and (b) the average distance it travels before decaying, as measured by an observer on Earth? (c) What distance would it travel if time dilation did not occur?

Solution

(a) If the proper lifetime of the pi meson is $\Delta t_p = 2.6 \times 10^{-8}$ s, the lifetime measured by an observer on Earth (moving at speed $v = 0.98\ c$ relative to the meson) is

$$\Delta t = \gamma\,\Delta t_p = \frac{\Delta t_p}{\sqrt{1-(v/c)^2}} = \frac{2.6 \times 10^{-8} \text{ s}}{\sqrt{1-(0.98)^2}} = 1.3 \times 10^{-7} \text{ s}$$ ◊

(b) The observer on Earth sees the meson travel at $0.98c$ for $\Delta t = 1.3 \times 10^{-7}$ s. Thus, the distance the meson travels during its lifetime according to this observer is

$$L_p = v(\Delta t) = 0.98(3.0 \times 10^8 \text{ m/s})(1.3 \times 10^{-7} \text{ s}) = 38 \text{ m}$$ ◊

(c) If time dilation did not occur, the Earth-based observer would have seen the meson travel at $0.98c$ for its proper lifetime of $\Delta t_p = 2.6 \times 10^{-8}$ s. Thus, the distance this observer would see the meson travel under these circumstances would be

$$L = v(\Delta t_p) = 0.98(3.0 \times 10^8 \text{ m/s})(2.6 \times 10^{-8} \text{ s}) = 7.6 \text{ m}$$ ◊

---

9.   The proper length of one spaceship is three times that of another. The two spaceships are traveling in the same direction and, while both are passing overhead, an Earth observer measures the two spaceships to have the same length. If the slower spaceship is moving with a speed of $0.350c$, determine the speed of the faster spaceship.

Solution

The faster object is the one that will appear to be contracted the most to the Earth-based observer. Since this observer measures the two moving ships to have the same length, the faster one must have the greater proper length, or $L_{pf} = 3L_{ps}$.

The contracted lengths that the Earth-based observer measures for the two ships are:

$$L_f = \frac{L_{pf}}{\gamma_f} = L_{pf}\sqrt{1-\left(v_f/c\right)^2}$$

and $$L_s = \frac{L_{ps}}{\gamma_s} = L_{ps}\sqrt{1-\left(v_s/c\right)^2} = L_{ps}\sqrt{1-(0.350)^2} = 0.937 L_{ps}$$

But, since this observer determines that $L_f = L_s$, and we have already determined that $L_{pf} = 3L_{ps}$, we find that

$$\left(3L_{ps}\right)\sqrt{1-\left(v_f/c\right)^2} = 0.937 L_{ps} \qquad \text{or} \qquad \sqrt{1-\left(v_f/c\right)^2} = 0.312$$

Thus,

$$v_f = c\sqrt{1-(0.312)^3} = 0.950\,c \qquad\qquad \lozenge$$

---

15.   An unstable particle at rest breaks up into two fragments of *unequal mass*. The mass of the lighter fragment is 2.50×10^{-28} kg and that of the heavier fragment is 1.67×10^{-27} kg. If the lighter fragment has a speed of 0.893c after the breakup, what is the speed of the heavier fragment?

Solution

The unstable particle is initially at rest, so the total momentum is zero before breakup. To conserve momentum, the total momentum after breakup must also be zero, meaning that the momenta of the two fragments must have equal magnitudes and opposite directions. If we label the lighter fragment 1 and the heavier one 2, equating magnitudes of momenta gives

$$p_2 = p_1 = \gamma_1 m_1 v_1 = \frac{\left(2.50 \times 10^{-28}\ \text{kg}\right)(0.893c)}{\sqrt{1-(0.893)^2}} = \left(4.96 \times 10^{-28}\ \text{kg}\right)c$$

We also know that $p_2 = \gamma_2 m_2 v_2$, so

$$\frac{\left(1.67 \times 10^{-27}\ \text{kg}\right)v_2}{\sqrt{1-\left(v_2/c\right)^2}} = \left(4.96 \times 10^{-28}\ \text{kg}\right)c$$

This reduces to

$$\left(v_2/c\right) = 0.297\sqrt{1-\left(v_2/c\right)^2}$$

Squaring both sides of this equation gives

$$\left(v/c\right)^2 = 0.088\ 2 - 0.088\ 2\left(v/c\right)^2$$

Thus,

$$v_2 = c\sqrt{\frac{0.088\ 2}{1+0.088\ 2}} = 0.285\,c \qquad \qquad \lozenge$$

---

19. What speed must a particle attain before its kinetic energy is double the value predicted by the nonrelativistic expression $KE = \frac{1}{2}mv^2$?

Solution

The nonrelativistic expression for kinetic energy is $KE_{nr} = \frac{1}{2}mv^2$, while the relativistic expression is $KE_r = E - E_R = (\gamma - 1)E_R$ where $E_R = mc^2$ and $\gamma = 1/\sqrt{1-(v/c)^2}$. Thus, if we are to have $KE_r = 2KE_{nr}$, it is necessary that

$$(\gamma - 1)mc^2 = mv^2 \qquad \text{or} \qquad \frac{1}{\sqrt{1-(v/c)^2}} - 1 = (v/c)^2$$

Rearranging and squaring this equation gives $1 = \left[1+(v/c)^2\right]^2\left[1-(v/c)^2\right]$.

Expanding and then simplifying this result yields

$$1 = \left[1+2(v/c)^2 + (v/c)^4\right]\left[1-(v/c)^2\right] = 1+(v/c)^2 - (v/c)^4 - (v/c)^6$$

and $\qquad 0 = -(v/c)^2\left[(v/c)^4 + (v/c)^2 - 1\right]$

Ignoring the trivial solution $v/c = 0$, we must have $(v/c)^4 + (v/c)^2 - 1$. If we let $x = (v/c)^2$, this becomes $x^2 + x - 1 = 0$, and the quadratic formula gives solutions of

$$x = \frac{-1 \pm \sqrt{1-4(1)(-1)}}{2} = \frac{-1 \pm \sqrt{5}}{2}$$

Since $x = (v/c)^2 \geq 0$, we must reject the negative solution and obtain

$$x = (v/c)^2 = \frac{-1+\sqrt{5}}{2} = 0.618$$

which yields

$$v = c\sqrt{0.618} = 0.786\,c \qquad \qquad \lozenge$$

23. Starting with the definitions of relativistic energy and momentum, show that $E^2 = p^2c^2 + m^2c^4$ (Equation (26.10)).

Solution

The definition of the total relativistic energy of a particle having mass m is $E = \gamma E_R$, where $\gamma = 1/\sqrt{1-(v/c)^2}$ and the rest energy is given by $E_R = mc^2$. The definition of the relativistic momentum is $p = \gamma mv$.

Therefore, we may write $E^2 = \gamma^2 m^2 c^4$ and $p^2 = \gamma^2 m^2 v^2$, or

$$\frac{E^2}{c^2} - p^2 = \gamma^2 m^2 c^2 - \gamma^2 m^2 v^2$$

$$= \gamma^2 m^2 \left(c^2 - v^2\right) = \gamma^2 m^2 c^2 \left(1 - v^2/c^2\right)$$

$$= \left(\frac{1}{1 - v^2/c^2}\right) m^2 c^2 \left(1 - v^2/c^2\right)$$

and we now have

$$\frac{E^2}{c^2} - p^2 = m^2 c^2$$

Multiplying through by c^2 and rearranging slightly yields the relativistic relation between total energy and momentum (Equation (26.10)).

$$E^2 - p^2 c^2 = m^2 c^4$$

or

$$E^2 = p^2 c^2 + m^2 c^4 = \left(pc\right)^2 + E_R^2 \qquad \Diamond$$

---

27. The rest energy of an electron is 0.511 MeV. The rest energy of a proton is 938 MeV. Assume both particles have kinetic energies of 2.00 MeV. Find the speed of (a) the electron and (b) the proton. (c) By how much does the speed of the electron exceed that of the proton? *Note:* perform the calculations in MeV; don't convert the energies to Joules. The answer is sensitive to rounding.

Solution

The relativistic expression for kinetic energy is

$$KE = E - E_R = \left(\gamma - 1\right)E_R$$

where $\gamma = 1/\sqrt{1-(v/c)^2}$ and the rest energy is $E_R = mc^2$. Thus, we may write

$$\gamma = 1 + KE/E_R \qquad \text{and} \qquad \gamma^2 = \frac{1}{1 - (v/c)^2} = \left(1 + KE/E_R\right)^2$$

Solving for the speed v of the particle then gives

$$\frac{1}{\left(1 + KE/E_R\right)^2} = 1 - \left(v/c\right)^2 \qquad \text{or} \qquad \left(v/c\right)^2 = 1 - \frac{1}{\left(1 + KE/E_R\right)^2}$$

Thus, $v = c \sqrt{1 - \dfrac{1}{\left(1 + KE/E_R\right)^2}}$

(a)   For an electron ($E_R = 0.511$ MeV) with $KE = 2.00$ MeV,

$$v_e = c \sqrt{1 - \frac{1}{\left(1 + 2.00 \text{ MeV}/0.511 \text{ MeV}\right)^2}} = 0.979\,c \qquad \lozenge$$

(b)   For a proton ($E_R = 938$ MeV) with $KE = 2.00$ MeV,

$$v_p = c \sqrt{1 - \frac{1}{\left(1 + 2.00 \text{ MeV}/938 \text{ MeV}\right)^2}} = 0.065\,2\,c \qquad \lozenge$$

(c) $v_e - v_p = 0.979\,c - 0.065\,2\,c = 0.914\,c$ $\qquad \lozenge$

---

31.   An astronaut wishes to visit the Andromeda galaxy, making a one-way trip that will take 30.0 years in the spaceship's frame of reference. Assume that the galaxy is 2.00 million light years away and that his speed is constant. (a) How fast must he travel relative to Earth? (b) What will be the kinetic energy of his spacecraft, which has a mass of 1.00×10^6 kg? (c) What is the cost of this energy if it is purchased at a typical consumer price for electric energy, 13.0 cents per kWh? The following approximation will prove useful:

$$\frac{1}{\sqrt{1+x}} \approx 1 - \frac{x}{2} \qquad \text{for} \quad x \ll 1$$

Solution

(a)   According to observers on Earth (moving at speed v relative to the astronaut), the time required for the trip will be

$$\Delta t = \gamma \left(\Delta t_p\right) = \frac{\Delta t_p}{\sqrt{1 - \left(v/c\right)^2}} = \frac{30.0 \text{ yr}}{\sqrt{1 - \left(v/c\right)^2}}$$

Since observers on Earth measure the distance traveled to be $d = 2.00 \times 10^6$ ly $= \left(2.00 \times 10^6 \text{ yr}\right)c$, the speed of the moving ship relative to Earth must be

$$v = \frac{d}{\Delta t} = \frac{\left(2.00 \times 10^6 \text{ yr}\right)c}{3.00 \text{ yr}\big/\sqrt{1 - \left(v/c\right)^2}}$$

which reduces to $\left(1.50 \times 10^{-5}\right)\dfrac{v}{c} = \sqrt{1 - \left(v/c\right)^2}$

Squaring both sides of this result and solving for v/c yields $\dfrac{v}{c} = \dfrac{1}{\sqrt{1 + 2.25 \times 10^{-10}}}$

and, using the approximation suggested in the problem statement, we find

$$\frac{v}{c} \approx 1 - \frac{2.25 \times 10^{-10}}{2} = 1 - 1.12 \times 10^{-10}$$

or $v = (1 - 1.12 \times 10^{-10})c$ ◊

(b) The kinetic energy of the spacecraft is $KE = E - E_R = (\gamma - 1)mc^2$, and

$$\gamma = \frac{1}{\sqrt{1 - (v/c)^2}} = \frac{1}{\sqrt{1 - (1 - 1.12 \times 10^{-10})^2}} = \frac{1}{\sqrt{2.24 \times 10^{-10}}} = 6.68 \times 10^4$$

Thus, $KE = (6.68 \times 10^4 - 1)(1.00 \times 10^6 \text{ kg})(3.00 \times 10^8 \text{ m/s})^2 = 6.01 \times 10^{27} \text{ J}$ ◊

(c) At a rate of 13.0 cents per kWh, the cost of providing this quantity of energy for the spacecraft is

$$cost = KE \times rate = \left[(6.01 \times 10^{27} \text{ J})\left(\frac{1 \text{ kWh}}{3.60 \times 10^6 \text{ J}} \right) \right] (\$0.13/\text{kWh}) = \$2.17 \times 10^{20}$$ ◊

---

35. The nonrelativistic expression for the momentum of a particle, $p_{nr} = mv$, can be used if $v \ll c$. For what speed does the use of this formula give an error in the momentum of (a) 1.00% and (b) 10.0%?

Solution

The relativistic expression for the momentum of a particle is

$$p_r = \frac{mv}{\sqrt{1 - (v/c)^2}} = \gamma mv$$

Thus, the error made when using the nonrelativistic expression is

$$\Delta p = p_r - p_{nr} = \gamma mv - mv = (\gamma - 1)mv$$

and the percentage error is

$$\% \text{ error} = \left(\frac{\Delta p}{p_r} \right) \times 100\% = \left[\frac{(\gamma - 1)mv}{\gamma mv} \right] \times 100\% = (1 - 1/\gamma) \times 100\% = \left(1 - \sqrt{1 - (v/c)^2} \right) \times 100\%$$

(a) If the error is 1.00%, we have

$$\left(1 - \sqrt{1 - (v/c)^2} \right) \times 100\% = 1.00\%   \text{ or }   \left(1 - \sqrt{1 - (v/c)^2} \right) = 0.010$$

Solving for the speed, v, gives $v = \sqrt{1 - (0.990)^2}\, c = 0.141c$ ◊

(b)   When the error is 10.0%, then

$$\left(1-\sqrt{1-(v/c)^2}\right) \times 100\% = 10.0\% \quad \text{or} \quad \left(1-\sqrt{1-(v/c)^2}\right) = 0.100$$

which yields $v = \sqrt{1-(0.900)^2}\ c = 0.436\,c$ ◊

---

40.   An interstellar space probe is launched from Earth. After a brief period of acceleration it moves with a constant velocity, 70.0% of the speed of light. Its nuclear-powered batteries supply the energy to keep its data transmitter active continuously. The batteries have a life-time of 15.0 years as measured in a rest frame. (a) How long do the batteries on the space probe last as measured by mission control on Earth? (b) How far is the probe from Earth when its batteries fail, as measured by mission control? (c) How far is the probe from Earth, as measured by its built-in trip odometer, when its batteries fail? (d) For what total time after launch are data received from the probe by mission control? Note that radio waves travel at the speed of light and fill the space between the probe and Earth at the time the battery fails.

Solution

(a)   The batteries have a lifetime of $\Delta t_p = 15.0$ yr in their own rest frame (the reference frame of the ship). Observers on Earth are moving at speed $v = 0.700\,c$ relative to this reference frame and hence measure a dilated lifetime of

$$\Delta t = \gamma(\Delta t_p) = \frac{\Delta t_p}{\sqrt{1-(v/c)^2}} = \frac{15.0\ \text{yr}}{\sqrt{1-(0.700)^2}} = 21.0\ \text{yr}$$ ◊

(b)   Mission control sees the ship travel outward at $v = 0.700\,c$ for 21.0 yr before battery failure. Thus, the distance they measure it to have traveled is

$$L_p = v(\Delta t) = (0.700\ c)(21.0\ \text{yr}) = (14.7\ \text{yr})c = 14.7\ \text{ly}$$ ◊

(c)   The probe sees Earth recede at $v = 0.700\,c$ for $\Delta t_p = 15.0$ yr (as measured by the probe's clock) before battery failure. The reading on the probe's trip odometer is

$$L = v(\Delta t_p) = (0.700\ c)(15.0\ \text{yr}) = (10.5\ \text{yr})c = 10.5\ \text{ly}$$ ◊

(d)   According to the Earth-based clock, mission control receives signals for 21.0 yr before the batteries fail. After the battery fails, mission control continues to receive signals until the information that left the ship at the instant the battery failed has traveled 14.7 ly (traveling at speed c) back to Earth. Thus, they receive signals for a total time of

$$\Delta t_{\text{total}} = \Delta t + (14.7\ \text{ly}/c) = 21.0\ \text{yr} + 14.7\ \text{yr} = 35.7\ \text{yr}$$ ◊

27

Quantum Physics

NOTES FROM SELECTED CHAPTER SECTIONS

27.1 Blackbody Radiation and Planck's Hypothesis

A **black body** is an ideal system that absorbs all radiant energy incident on it. The nature of the radiation emitted by a black body depends on its temperature. The spectral distribution of blackbody radiation (thermal radiation) at various temperatures is sketched in the figure below.

As the temperature increases, the total radiation emitted (area under the curve) increases, while the peak of the distribution shifts to shorter wavelengths. **Stefan's law** describes the total power radiated as a function of temperature. The shift to shorter wavelengths is consistent with the **Wien displacement law.** See Equation (27.1). Classical theories failed to explain the spectrum of blackbody radiation as observed experimentally. An empirical formula, proposed by Max Planck, is consistent with the observed distribution at all wavelengths. Planck made two basic assumptions:

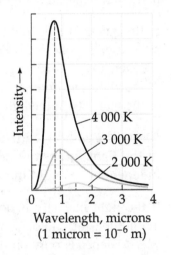

Blackbody Radiation Spectrum

(1) Blackbody radiation is caused by submicroscopic charged oscillators called **resonators.**

(2) The resonators could only have **discrete energies (quantum states)** given by $E_n = nhf$, where f is the oscillator frequency, n is a quantum number ($n = 1, 2, 3, \ldots$), and h is Planck's constant.

Subsequent developments showed that the quantum concept was necessary in order to explain several phenomena at the atomic level, including the photoelectric effect, the Compton effect, and atomic spectra.

27.2   The Photoelectric Effect and the Particle Theory of Light
When light, of sufficiently high frequency, is
incident on certain metallic surfaces, electrons
will be emitted from the surfaces. This is called
the **photoelectric effect,** discovered by Hertz.

Several features of the photoelectric effect cannot
be explained with classical physics or with the
wave theory of light. However, each of these features can be explained and understood on the basis
of the photon theory of light. These features and
their explanations include:

Metal

Photoelectron Emission

1. **No electrons are emitted if the frequency of the incident light is below a value
    called the cutoff frequency, f_c.** The value of the cutoff frequency depends on the
    particular material being illuminated.

    In the quantum model, the energy of electromagnetic radiation is assumed to occur in
    packets called photons. *Photons in the incident light interact with individual electrons.*
    If the energy of an incoming photon is not equal to or greater than the work function, ϕ,
    of the photosensitive surface, no electrons will be ejected from the surface, regardless
    of the intensity of the light.

2. **Above the cutoff frequency, the number of photoelectrons emitted is proportional
    to the light intensity and the maximum kinetic energy of the photoelectrons is
    independent of light intensity.**

    If the light intensity is doubled, the number of photons is doubled, which doubles
    the number of photoelectrons emitted. However, their kinetic energy, which equals
    $hf - \phi$, depends only on the light frequency and the work function, not on the light
    intensity.

3. **The maximum kinetic energy of the photoelectrons increases with increasing
    light frequency.**

    The fact that KE_{max} increases with increasing frequency is easily understood from
    Equation (27.6), $KE_{max} = hf - \phi$.

4. **Electrons are emitted from the surface almost instantaneously (less than 10^{-9} s
    after the surface is illuminated), even at low light intensities.**

    Classically, one would expect that the electrons would require some time to absorb
    the incident radiation before they acquire enough kinetic energy to escape from the
    metal. The fact that the electrons are emitted almost instantaneously is consistent
    with the photon theory of light; there is a one-to-one interaction between photons and
    electrons.

27.3   X-Rays

X-rays are a part of the electromagnetic spectrum, characterized by frequencies higher than those of ultraviolet radiation with wavelengths in the 10 to 10^{-4} nm range. They are produced when a metal (and in some cases non-metal) target is struck by high speed electrons. The spectrum of radiation emitted by an x-ray tube has two distinct components:

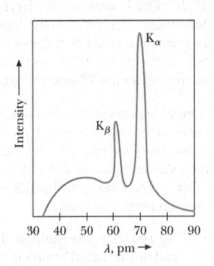

> **Bremsstrahlung** — a continuous broad spectrum of high-energy photons produced by accelerating electrons as they pass close to large positive nuclei in the metal target.

> **Characteristic lines** — a series of sharp lines
characteristic of the target material which appear superimposed on the continuous spectrum as the accelerating voltage exceeds the threshold value. These lines represent photons emitted as electrons in atoms of the target material undergo rearrangements following collisions with the incoming electrons.

27.5   The Compton Effect

The Compton effect involves the scattering of a photon by an electron. The scattered photon undergoes a change in wavelength $\Delta\lambda$, called the **Compton shift.** An increase in wavelength for a scattered photon can be explained on the basis of a scattering event in which a portion of the energy of an incident ("bombarding") photon is transferred to an electron ("target"). *The total energy and the total momentum of the photon-electron pair are conserved during the scattering event.*

The "bombarding-photon" and "target-electron" model predicts wavelength shifts which are in excellent agreement with experimental results.

The figure at the right shows the Compton shift for x-rays scattered at 90° from graphite. In this case, the Compton shift is $\Delta\lambda = 0.002\,43$ nm, and λ_0 is the wavelength of photons in the incident x-ray beam.

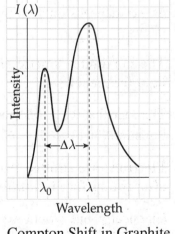

Compton Shift in Graphite

27.6   The Dual Nature of Light and Matter

The results of some experiments (i.e., photoelectric emission of electrons and Compton scattering) can only be explained on the basis of the photon model of light. Other phenomena (i.e., interference and diffraction of light) are consistent with the wave model. *The photon theory and the wave theory complement each other: light exhibits both wave and particle characteristics.*

De Broglie postulated that a particle in motion has wave properties and a corresponding wavelength inversely proportional to the particle's momentum. The wave nature of matter was first confirmed (Davisson and Germer) by scattering electrons from crystal targets.

27.7 The Wave Function

In the Schrödinger description of the manner in which matter waves change in space and time, each particle is represented by a wave function. *The probability of finding a particle at a specific location at some instant is given by evaluating the square of the wave function at that location and time.*

27.8 The Uncertainty Principle

It is impossible to make exact simultaneous measurements of a particle's position and linear momentum or to measure exactly the energy of a particle in an arbitrarily small time interval. These limitations are described by Equations (27.16) and (27.17).

EQUATIONS AND CONCEPTS

The **Wien displacement law** accurately describes the manner in which the peak wavelength in the spectrum of a blackbody radiator depends on temperature. *As the temperature increases, the intensity of the radiation (area under the intensity vs. wavelength curve shown in Section 27.1) increases, while the peak of the distribution shifts to shorter wavelengths.*

$$\lambda_{max} T = 0.289\,8 \times 10^{-2} \text{ m} \cdot \text{K} \qquad (27.1)$$

Discrete energy values of an atomic oscillator are determined by a quantum number, n. Each discrete energy value corresponds to a quantum state. **Planck's constant** (h) is a fundamental constant of nature.

$$E_n = nhf \qquad (27.2)$$
$$(n = 1, 2, 3, \dots)$$

$$h = 6.626 \times 10^{-34} \text{ J} \cdot \text{s} \qquad (27.3)$$

The **energy of a photon** corresponds to the energy difference between initial and final quantum states. *An oscillator emits or absorbs energy only when there is a transition between quantum states.*

$$E = hf \qquad (27.5)$$

The **maximum kinetic energy of an ejected photoelectron** (typically a few eV) can be expressed in terms of the stopping potential (ΔV_s) or in terms of the work function of the metal (ϕ) and the frequency of the incident photon.

$$KE_{max} = e\Delta V_s \tag{27.4}$$

$$KE_{max} = hf - \phi \tag{27.6}$$

The **cut-off wavelength** is the maximum wavelength (corresponding to the minimum frequency) of an incident photon that will result in photoemission regardless of the intensity of the incident light.

$$\lambda_c = \frac{hc}{\phi} \tag{27.7}$$

$$hc = 1\,240 \text{ eV} \cdot \text{nm}$$

The **minimum wavelength x-ray** radiation, for a given accelerating voltage, is produced when an electron is completely stopped in a single collision.

$$\lambda_{min} = \frac{hc}{e\Delta V} \tag{27.9}$$

Bragg's law states the condition for constructive interference of x-rays diffracted by a crystal. The crystal planes are separated by a distance d; x-rays are incident at an angle θ to the crystal plane and are reflected at the same angle.

$$2d\sin\theta = m\lambda \tag{27.10}$$
$$(m = 1, 2, 3, \ldots)$$

The **Compton shift** is the change in wavelength of a photon when scattered from an electron. *The scattered photon makes an angle θ with the direction of the incident photon.*

$$\Delta\lambda = \lambda - \lambda_o = \frac{h}{m_e c}(1 - \cos\theta) \tag{27.11}$$

$$h/m_e c = \text{Compton wavelength}$$
$$= 0.002\,43 \text{ nm}$$

The **momentum and wavelength** of a photon are related as shown in Equation (27.13). This equation is obtained by combining two equations involving the energy of a photon: $p = E/c$ and $E = hc/\lambda$.

$$p = \frac{E}{c} = \frac{h}{\lambda} \tag{27.13}$$

The **de Broglie wavelength** of a particle is inversely proportional to the momentum of the particle. *The waves associated with material particles are called **matter waves** with frequencies given by Equation (27.15).*

$$\lambda = \frac{h}{mv} \qquad (27.14)$$

$$f = \frac{E}{h} \qquad (27.15)$$

The **Heisenberg uncertainty principle** can be stated in two forms:

- **Simultaneous measurements of position and momentum** with respective uncertainties Δx and Δp_x.

$$\Delta x \, \Delta p_x \geq \frac{h}{4\pi} \qquad (27.16)$$

- **Simultaneous measurements of energy and lifetime** with uncertainties ΔE and Δt.

$$\Delta E \, \Delta t \geq \frac{h}{4\pi} \qquad (27.17)$$

REVIEW CHECKLIST

- State the assumptions made by Planck in order to explain the shape of the Intensity vs. Wavelength curve for a blackbody radiator. (Section 27.1)

- Describe the manner in which the Intensity vs. Wavelength curve for a blackbody radiator changes as the temperature of the radiator increases. (Section 27.1)

- Describe the Einstein model for the photoelectric effect and the predictions of the fundamental photoelectric effect equation for the maximum kinetic energy of photoelectrons. (Section 27.2)

- Describe the Compton effect (the scattering of x-rays by electrons) and be able to use the formula for the Compton shift equation. Recognize that the Compton effect can only be explained using the photon concept. (Section 27.5)

- Use the de Broglie equation to calculate the wavelength associated with moving particles. (Section 27.6)

- Identify experimental evidence for the wave and particle nature of light and also for the wave nature of particles. (Section 27.6)

- Discuss the manner in which the uncertainty principle makes possible a better understanding of the dual wave-particle nature of light and matter. (Section 27.8)

SOLUTIONS TO SELECTED END-OF-CHAPTER PROBLEMS

1. (a) What is the surface temperature of Betelgeuse, a red giant star in the constellation of Orion, which radiates with a peak wavelength of about 970 nm? (b) Rigel, a bluish-white star in Orion, radiates with a peak wavelength of 145 nm. Find the temperature of Rigel's surface.

Solution

If we assume that the surface of a star approximates a blackbody radiator, the absolute temperature of the surface and the peak wavelength of the radiation from that star are related by Wien's displacement law,

$$\lambda_{max} T = 0.289\ 8 \times 10^{-2}\ \text{m} \cdot \text{K}$$

(a) For Betelgeuse, a red giant star, $\lambda_{max} = 970$ nm. Thus, the surface temperature is

$$T = \frac{0.289\ 8 \times 10^{-2}\ \text{m} \cdot \text{K}}{\lambda_{max}} = \frac{0.289\ 8 \times 10^{-2}\ \text{m} \cdot \text{K}}{970 \times 10^{-9}\ \text{m}} = 2.99 \times 10^3\ \text{K} \approx 3\ 000\ \text{K} \qquad \lozenge$$

(b) The peak wavelength from Rigel is 145 nm, giving a surface temperature of

$$T = \frac{0.289\ 8 \times 10^{-2}\ \text{m} \cdot \text{K}}{\lambda_{max}} = \frac{0.289\ 8 \times 10^{-2}\ \text{m} \cdot \text{K}}{145 \times 10^{-9}\ \text{m}} = 2.00 \times 10^4\ \text{K} \approx 20\ 000\ \text{K} \qquad \lozenge$$

---

6. Suppose a star with radius 8.50×10^8 m has a peak of 685 nm in the spectrum of its emitted radiation. (a) Find the energy of a photon with this wavelength. (b) What is surface temperature of the star? (c) At what rate is energy emitted from the star in the form of radiation? (d) Using the answer to (a), estimate the rate at which photons leave the surface of the star.

Solution

(a) The energy of a photon of frequency f and wavelength $\lambda = c/f$ is $E = hf = hc/\lambda$. The energy per 685-nm photon is then

$$E_{peak} = \frac{hc}{\lambda_{max}} = \frac{\left(6.63 \times 10^{-34}\ \text{J} \cdot \text{s}\right)\left(3.00 \times 10^8\ \text{m/s}\right)}{685 \times 10^{-9}\ \text{m}} = 2.90 \times 10^{-19}\ \text{J/photon} \qquad \lozenge$$

(b) Assuming the star closely approximates an ideal radiator or blackbody, the absolute temperature of its surface and the wavelength of its peak radiation are related by Wien's displacement law, $\lambda_{max} T = 0.289\ 8 \times 10^{-2}$ m $\cdot$ K. The surface temperature must be

$$T = \frac{0.289\ 8 \times 10^{-2}\ \text{m} \cdot \text{K}}{\lambda_{max}} = \frac{0.289\ 8 \times 10^{-2}\ \text{m} \cdot \text{K}}{685 \times 10^{-9}\ \text{m}} = 4.23 \times 10^3\ \text{K} \qquad \lozenge$$

(c) Stefan's law (see Chapter 11 in the textbook) states that the rate at which an object radiates energy is proportional to the fourth power of its absolute temperature, or $P = \sigma A e T^4$ where $\sigma = 5.669\ 6 \times 10^{-8}$ W/m$^2 \cdot$K^4, A is the surface area of the object, and e is the emissivity. For an ideal radiator, the emissivity is $e = 1$. Assuming the star to be spherical, $A = 4\pi r^2$, and the radiated power must be

$$P = \left(5.669\ 6 \times 10^{-8}\ \text{W/m}^2 \cdot \text{K}^4\right) 4\pi \left(8.50 \times 10^8\ \text{m}\right)^2 (1)\left(4.23 \times 10^3\ \text{K}\right)^4 = 1.65 \times 10^{26}\ \text{W} \qquad \lozenge$$

(d) If we assume that the average energy of the photons emitted by the star equals the energy of photons at the peak of the radiation distribution, the estimate of the number of photons emitted per second would be

$$\frac{\Delta n}{\Delta t} = \frac{P}{E_{\text{peak}}} = \frac{1.65 \times 10^{26}\ \text{J/s}}{2.90 \times 10^{-19}\ \text{J/photon}} = 5.69 \times 10^{44}\ \text{photons/s} \qquad \lozenge$$

---

9. When light of wavelength 350 nm falls on a potassium surface, electrons having a maximum kinetic energy of 1.31 eV are emitted. Find (a) the work function of potassium, (b) the cutoff wavelength, and (c) the frequency corresponding to the cutoff wavelength.

Solution

(a) Einstein's photoelectric effect equation is $KE_{\text{max}} = E_{\text{photon}} - \phi$, where KE_{max} is the maximum kinetic energy of the emitted electrons, E_{photon} is the energy of the photons incident on the surface, and ϕ is the work function of the material making up the surface. If the wavelength of the incident light is $\lambda = 350$ nm,

$$E_{\text{photon}} = hf = \frac{hc}{\lambda} = \frac{\left(6.63 \times 10^{-34}\ \text{J} \cdot \text{s}\right)\left(3.00 \times 10^8\ \text{m/s}\right)}{350 \times 10^{-9}\ \text{m}} = 5.68 \times 10^{-19}\ \text{J}$$

or $\quad E_{\text{photon}} = \left(5.68 \times 10^{-19}\ \text{J}\right)\left(\dfrac{1.00\ \text{eV}}{1.60 \times 10^{-19}\ \text{J}}\right) = 3.55\ \text{eV}$

The observed maximum kinetic energy of the emitted electrons is $KE_{\text{max}} = 1.31$ eV. Thus, the work function of the potassium surface is

$$\phi = E_{\text{photon}} - KE_{\text{max}} = 3.55\ \text{eV} - 1.31\ \text{eV} = 2.24\ \text{eV} \qquad \lozenge$$

(b)   The cutoff wavelength is the wavelength of the lowest energy photons capable of freeing electrons from the surface. At the cutoff wavelength $(\lambda = \lambda_c)$, the freed electrons leave the surface with zero kinetic energy $\left(\text{that is, } KE_{max} = 0\right)$.

From the photoelectric effect equation, $0 = \left(E_{photon}\right)_{min} - \phi$

or $\left(E_{photon}\right)_{min} = \dfrac{hc}{\lambda_c} = \phi$ and $\lambda_c = \dfrac{hc}{\phi}$

Thus, $\lambda_c = \dfrac{\left(6.63 \times 10^{-34} \text{ J} \cdot \text{s}\right)\left(3.00 \times 10^{8} \text{ m/s}\right)}{2.24 \text{ eV}\left(1.60 \times 10^{-19} \text{ J}/1.00 \text{ eV}\right)} = 5.55 \times 10^{-7} \text{ m} = 555 \text{ nm}$ ◊

(c)   The frequency corresponding to the cutoff wavelength is

$$f_c = \frac{c}{\lambda_c} = \frac{3.00 \times 10^{8} \text{ m/s}}{5.55 \times 10^{-7} \text{ m}} = 5.41 \times 10^{14} \text{ Hz}$$ ◊

---

13.   When light of wavelength 254 nm falls on cesium, the required stopping potential is 3.00 V. If light of wavelength 436 nm is used, the stopping potential is 0.900 V. Use this information to plot a graph like that shown in Figure 27.6, and from the graph determine the cutoff frequency for cesium and its work function.

Solution

The photoelectric effect equation,

$$KE_{max} = E_{photon} - \phi = hf - \phi$$

is a linear equation relating the maximum kinetic energy of the ejected electrons to the frequency of the incident radiation. Thus, the graph of KE_{max} versus f should be a straight line with slope h and vertical intercept $-\phi$.

We may express KE_{max} in terms of the stopping potential as $KE_{max} = V_s e$. Therefore, we have

$$KE_{max} = \left(3.00 \text{ V}\right)e = 3.00 \text{ eV} \quad \text{when} \quad f = \frac{c}{\lambda} = \frac{3.00 \times 10^{8} \text{ m/s}}{254 \times 10^{-9} \text{ m}} = 11.8 \times 10^{14} \text{ Hz}$$

and

$$KE_{max} = \left(0.900 \text{ V}\right)e = 0.900 \text{ eV} \quad \text{when} \quad f = \frac{c}{\lambda} = \frac{3.00 \times 10^{8} \text{ m/s}}{436 \times 10^{-9} \text{ m}} = 6.88 \times 10^{14} \text{ Hz}$$

Your graph should be similar to the sketch given at the right. At the cutoff frequency (the minimum frequency photon capable of freeing electrons from the surface), $KE_{max} = 0$.

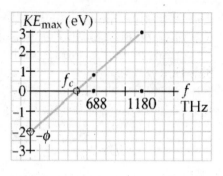

Thus, the horizontal intercept on the graph gives the cutoff frequency. The vertical intercept of the graph is equal to the negative of the work function. From your graph, you should find these values to be approximately

$$f_c = 480 \text{ THz} = 4.8 \times 10^{14} \text{ Hz} \quad \text{and} \quad \phi = 2.0 \text{ eV} \qquad \Diamond$$

19. Potassium iodide has an interplanar spacing of $d = 0.296$ nm. A monochromatic x-ray beam shows a first-order diffraction maximum when the grazing angle is $7.6°$. Calculate the x-ray wavelength.

Solution

The sketch at the right shows x-rays reflecting from adjacent planes or layers of atoms in a crystal having interplanar spacing d. When the difference in the path lengths, $\delta = 2d \sin\theta$, of the x-rays reflecting from adjacent layers equals an integral number of wavelengths, strong reflections (maxima due to constructive interference) will be observed. This is summarized in the equation known as Bragg's law:

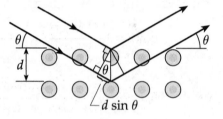

$$2d \sin\theta = m\lambda \qquad \text{where} \qquad m = 1, 2, 3, \ldots$$

Thus, if the first-order ($m = 1$) maximum occurs at a grazing angle of $\theta = 7.6°$ when x-rays strike a crystal having interplanar spacing $d = 0.296$ nm, the wavelength of the x-rays is

$$\lambda = \frac{2d \sin\theta}{m} = \frac{2(0.296 \text{ nm})\sin(7.6°)}{1} = 0.078 \text{ nm} \qquad \Diamond$$

23.   A 0.001 6-nm photon scatters from a free electron. For what (photon) scattering angle will the recoiling electron and scattered photon have the same kinetic energy?

Solution

Before the scattering event, the energy of the free electron was just its rest energy E_R. After the event, the total energy of the electron is

$$E = KE + E_R$$

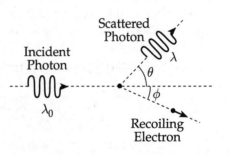

The energy of the incident photon is $E_0 = hc/\lambda_0$, while that of the scattered photon is $E' = hc/\lambda$. Conservation of energy then gives

$$E_0 + E_R = E' + KE + E_R \quad \text{or} \quad E_0 = E' + KE$$

Thus, if the kinetic energy of the recoiling electron is the same as the energy of the scattered photon $(KE = E')$, we have $E_0 = E' + E' = 2E'$, giving

$$\frac{hc}{\lambda_0} = 2\left(\frac{hc}{\lambda}\right) \quad \text{or} \quad \lambda = 2\lambda_0$$

The Compton shift formula, $\Delta\lambda = \lambda - \lambda_0 = \dfrac{h}{m_e c}(1 - \cos\theta)$, then gives the photon scattering angle as

$$\cos\theta = 1 - \left(\frac{m_e c}{h}\right)\Delta\lambda = 1 - \left(\frac{m_e c}{h}\right)(2\lambda_0 - \lambda_0) = 1 - \left(\frac{m_e c}{h}\right)\lambda_0$$

or $$\theta = \cos^{-1}\left[1 - \frac{(9.11 \times 10^{-31}\ \text{kg})(3.00 \times 10^8\ \text{m/s})}{6.63 \times 10^{-34}\ \text{J}\cdot\text{s}}(0.001\,6 \times 10^{-9}\ \text{m})\right] = 70° \qquad \lozenge$$

---

29.   De Broglie postulated that the relationship $\lambda = h/p$ is valid for relativistic particles. What is the de Broglie wavelength for a (relativistic) electron whose kinetic energy is 3.00 MeV?

Solution

The kinetic energy of a 3.00-MeV electron is not small in comparison to the rest energy $(E_R = 0.511\ \text{MeV})$ of an electron. Thus, a relativistic calculation for the momentum of this particle is required.

The relativistic relation between the total energy of a particle and its momentum is

$$E^2 = p^2 c^2 + E_R^2$$

where the total energy is $E = KE + E_R$. Thus, the momentum of a relativistic particle may be expressed as

$$p = \frac{\sqrt{E^2 - E_R^2}}{c} = \frac{\sqrt{(KE + E_R)^2 - E_R^2}}{c}$$

and its de Broglie wavelength is

$$\lambda = \frac{h}{p} = \frac{hc}{\sqrt{\left(KE + E_R\right)^2 - E_R^2}}$$

An electron having a kinetic energy of $KE = 3.00$ MeV has a de Broglie wavelength of

$$\lambda = \frac{\left(6.63 \times 10^{-34} \text{ J} \cdot \text{s}\right)\left(3.00 \times 10^8 \text{ m/s}\right)}{\sqrt{\left(3.00 \text{ MeV} + 0.511 \text{ MeV}\right)^2 - \left(0.511 \text{ MeV}\right)^2}} \left(\frac{1 \text{ MeV}}{1.60 \times 10^{-13} \text{ J}}\right) = 3.58 \times 10^{-13} \text{ m} \qquad \lozenge$$

33. In the ground state of hydrogen, the uncertainty in the position of the electron is roughly 0.10 nm. If the speed of the electron is approximately the same as the uncertainty in its speed, about how fast is it moving?

Solution

One form of the uncertainty principle states that if the position of a particle is measured with precision Δx and a simultaneous measurement of the linear momentum is made with precision Δp_x, the product of the two uncertainties can never be less than $h/4\pi$, or $\Delta x \, \Delta p_x \geq h/4\pi$.

Thus, if we assume that the momentum of the electron is not less than the uncertainty in the momentum, the minimum momentum of the electron is

$$p_{min} = \left(\Delta p_x\right)_{min} = \frac{h}{4\pi \, \Delta x} = \frac{6.63 \times 10^{-34} \text{ J} \cdot \text{s}}{4\pi \left(0.10 \text{ nm}\right)\left(10^{-9} \text{ m/1 nm}\right)} = 5.3 \times 10^{-25} \text{ kg} \cdot \text{m/s}$$

We shall assume that this electron is not relativistic and try a classical calculation for the minimum speed of the particle. Then,

$$v_{min} = \frac{p_{min}}{m_e} = \frac{5.3 \times 10^{-25} \text{ kg} \cdot \text{m/s}}{9.11 \times 10^{-31} \text{ kg}} = 5.8 \times 10^5 \text{ m/s} \quad \text{or} \quad v_{min} \quad 10^6 \text{ m/s} \qquad \lozenge$$

Observe that the computed speed is 10^6 m/s $\ll c$, and our assumption that a relativistic calculation would not be necessary is seen to be valid.

37. The average lifetime of a muon is about $2 \ \mu\text{s}$. Estimate the minimum uncertainty in the energy of a muon.

Solution

A second form of the uncertainty principle sets a limit on the accuracy with which the energy of a system E can be measured in a finite time interval Δt. This form states that the product of the uncertainty in the energy and the duration of the interval over which that energy is measured can never be less than $h/4\pi$, or $\Delta E \, \Delta t \geq h/4\pi$.

If, on average, a muon only exists for a lifetime of $\tau = 2 \ \mu\text{s}$, the maximum duration of the time interval over which its energy could be monitored and measured is $\Delta t_{\text{max}} = \tau = 2 \ \mu\text{s}$. Thus, the minimum uncertainty in the measurement of its energy is

$$\Delta E_{\text{min}} = \frac{h}{4\pi \, \Delta t_{\text{max}}} = \frac{6.63 \times 10^{-34} \ \text{J} \cdot \text{s}}{4\pi \left(2 \times 10^{-6} \ \text{s}\right)} = 3 \times 10^{-29} \ \text{J} \qquad \Diamond$$

40. Find the speed of an electron having a de Broglie wavelength equal to its Compton wavelength. (*Hint:* This electron is relativistic.)

Solution

The de Broglie wavelength of a particle is $\lambda = h/p$, where p is the linear momentum of the particle, and the Compton wavelength for a particle is $\lambda_C = h/mc$, where m is the mass of the particle. Thus, if the de Broglie wavelength for an electron is equal to its Compton wavelength, it is necessary that $h/p = h/m_e c$, or the linear momentum of the electron must be $p = m_e c$.

The relativistic relation between the total energy of a particle and its momentum is $E^2 = (pc)^2 + E_R^2$, where the rest energy is $E_R = mc^2$, and the total energy is $E = \gamma E_R = mc^2 / \sqrt{1 - (v/c)^2}$. Applying this to our electron having $p = m_e c$ gives

$$\frac{m_e^2 c^4}{1 - (v/c)^2} = m_e^2 c^4 + m_e^2 c^4 = 2m_e^2 c^4$$

or

$$\frac{1}{2} = 1 - (v/c)^2$$

Solving for the speed of the electron then gives

$$v = c\sqrt{1 - \frac{1}{2}} = \frac{c}{\sqrt{2}} \quad \text{or} \quad v = c\sqrt{2}/2 \qquad \Diamond$$

45. Photons of wavelength 450 nm are incident on a metal. The most energetic electrons ejected from the metal are bent into a circular arc of radius 20.0 cm by a magnetic field with a magnitude of 2.00×10^{-5} T. What is the work function of the metal?

Solution

When electrons move perpendicularly to a magnetic field, they are deflected into a circular path with the required centripetal acceleration produced by the magnetic force acting on them. Thus,

$$m_e \left(\frac{v^2}{r} \right) = evB \quad \text{or} \quad v = \frac{eBr}{m_e}$$

The kinetic energy of the most energetic electrons ejected from the metal is

$$KE_{max} = \frac{1}{2} m_e v^2 = \frac{e^2 B^2 r^2}{2m_e} = \frac{\left(1.60 \times 10^{-19} \text{ C} \right)^2 \left(2.00 \times 10^{-5} \text{ T} \right)^2 \left(0.200 \text{ m} \right)^2}{2 \left(9.11 \times 10^{-31} \text{ kg} \right)}$$

or $\qquad KE_{max} = 2.25 \times 10^{-19}$ J

The energy of the photons incident on the metal surface is

$$E_{photon} = hf = \frac{hc}{\lambda} = \frac{\left(6.63 \times 10^{-34} \text{ J} \cdot \text{s} \right) \left(3.00 \times 10^8 \text{ m/s} \right)}{450 \times 10^{-9} \text{ m}} = 4.42 \times 10^{-19} \text{ J}$$

Einstein's photoelectric effect equation, $KE_{max} = E_{photon} - \phi$, then gives the work function of the metal as

$$\phi = E_{photon} - KE_{max} = 4.42 \times 10^{-19} \text{ J} - 2.25 \times 10^{-19} \text{ J} = 2.17 \times 10^{-19} \text{ J} \qquad \Diamond$$

or

$$\phi = 2.17 \times 10^{-19} \text{ J} \left(\frac{1 \text{ eV}}{1.60 \times 10^{-19} \text{ J}} \right) = 1.36 \text{ eV} \qquad \Diamond$$

49. How fast must an electron be moving if all its kinetic energy is lost to a single x-ray photon (a) at the high end of the x-ray electromagnetic spectrum with a wavelength of 1.00×10^{-8} m; (b) at the low end of the x-ray electromagnetic spectrum with a wavelength of 1.00×10^{-13} m?

Solution

Photons with a wavelength of $\lambda = 1.00 \times 10^{-8}$ m have energy

$$E_{photon} = hf = \frac{hc}{\lambda} = \frac{\left(6.63 \times 10^{-34} \text{ J} \cdot \text{s}\right)\left(3.00 \times 10^{8} \text{ m/s}\right)}{1.00 \times 10^{-8} \text{ m}} = 1.99 \times 10^{-17} \text{ J}$$

and

$$E_{photon} = 1.99 \times 10^{-17} \text{ J}\left(\frac{1 \text{ eV}}{1.60 \times 10^{-19} \text{ J}}\right) = 124 \text{ eV}$$

Those with $\lambda' = 10^{-5}\lambda = 1.00 \times 10^{-13}$ m have energy $E'_{photon} = 10^{5} E_{photon} = 12.4$ MeV.

(a) If an electron (with rest energy $E_R = 0.511$ MeV) has kinetic energy $KE = E_{photon} = 124$ eV, then $KE \ll E_R$ and the electron is non-relativistic. The speed of the electron is then

$$v = \sqrt{\frac{2 \, KE}{m_e}} = \sqrt{\frac{2\left(1.99 \times 10^{-17} \text{ J}\right)}{9.11 \times 10^{-31} \text{ kg}}} = 6.61 \times 10^{6} \text{ m/s} = 0.022 \, 0c \qquad \lozenge$$

(b) An electron having $KE = E'_{photon} = 12.4$ MeV is highly relativistic, and its total energy is $E = \gamma E_R = KE + E_R$, which gives

$$\gamma = 1 + \frac{KE}{E_R} = 1 + \frac{12.4 \text{ MeV}}{0.511 \text{ MeV}} = 25.3$$

Since $\gamma = 1/\sqrt{1 - (v/c)^2}$, the speed of the electron must be

$$v = c\sqrt{1 - \frac{1}{\gamma^2}} = c\sqrt{1 - \frac{1}{(25.3)^2}} = 0.999 \, 2 \, c \qquad \lozenge$$

28

Atomic Physics

NOTES FROM SELECTED CHAPTER SECTIONS

28.2   Atomic Spectra
28.3   The Bohr Theory of Hydrogen
The basic postulates of the Bohr model of the hydrogen atom are:

- The **electron moves in circular orbits** about the nucleus under the influence of the Coulomb force of attraction between the electron and the positively charged proton in the nucleus.

- The **electron can exist only in specific stationary states** (orbits). When the electron is in one of its allowed orbits, it does not emit energy by radiation.

- The **atom radiates energy only when the electron makes a transition** ("jumps") from one allowed stationary orbit to another less energetic state. This postulate states that the energy given off by an atom is carried away by a photon of energy *hf*.

- The **orbital angular momentum of the electron about the nucleus is quantized** in units of $n(h/2\pi)$, where *h* is Planck's constant. *This condition determines the radii of the allowed orbits.*

There are several important reasons to understand the behavior of the hydrogen atom as an atomic system:

- Much of what is learned about the hydrogen atom with its single electron can be extended to such single-electron ions as He^+ and Li^{2+}, which are hydrogen-like in their atomic structure.

- The hydrogen atom is an ideal system for performing precise tests of theory against experiment and for improving our overall understanding of atomic structure.

- The quantum numbers used to characterize the allowed states of hydrogen can be used to describe the allowed states of more complex atoms. This enables us to understand the periodic table of the elements.

- The basic ideas about atomic structure must be well understood before we attempt to deal with the complexities of molecular structures and the electronic structure of solids.

28.4 Quantum Mechanics and the Hydrogen Atom
The possible stationary energy states of an electron in an atom are determined by the values of four quantum numbers.

- **Principal quantum number (n) with integer values from 1 to ∞**

  The principal quantum number follows from the concept of quantization of angular momentum. Electron energy states with the same principal quantum number form a **shell.** Shells are identified by the letters K, L, M . . . corresponding to $n = 1, 2, 3$ . . ..

- **Orbital quantum number (ℓ) with integer values from 0 to $n - 1$**

  An electron in a given allowed energy state may move in different elliptical orbits determined by the value of ℓ. For each value of n there are n possible orbits corresponding to different values of ℓ. Energy states with given values of n and ℓ form a **subshell.** Subshells are identified by the letters s, p, d, f, . . . corresponding to $\ell = 0, 1, 2, 3, ...$ The maximum number of electrons allowed in any subshell is $2(2\ell + 1)$.

- **Orbital magnetic quantum number (m_ℓ) with integer values from $-\ell$ to $+\ell$**

  The orbital magnetic quantum number accounts for the observed **Zeeman effect:** when a gas is placed in an external magnetic field, single spectral lines are split into several lines.

- **Spin magnetic quantum number (m_s) with values of $-1/2$ and $+1/2$**

  The **spin magnetic quantum number** m_s accounts for the two closely spaced energy states in spectral lines ("**doublets**") corresponding to the two possible orientations of electron spin ($+1/2$ is "up" spin and $-1/2$ is "down" spin).

28.5 The Exclusion Principle and the Periodic Table
At this point you might find it useful to review the description of the quantum numbers in Section 28.4 of the textbook.

The Pauli exclusion principle states that no two electrons in an atom can exist in identical quantum states. This means that no two electrons in a given atom can have exactly the same values for the set of quantum numbers n, ℓ, m_ℓ, and m_s.

The order in which electrons fill subshells and the application of the exclusion principle are illustrated in the following tables in your textbook:

Table 28.3, Number of Electrons in Filled Subshells and Shells

Table 28.4, Electronic Configuration of Some Elements

28.6 Characteristic X-Rays

The x-ray spectrum of a metal target consists of a broad continuous spectrum on which are superimposed a series of sharp lines. **Characteristic x-rays** are emitted by atoms when an electron undergoes a transition from an outer shell into an electron vacancy in one of the inner shells. Transitions into a vacant state in the K shell give rise to the K series of spectral lines, transitions into a vacant state in the L shell create the L series of lines, and so on. *Lines within a series are given a notation to designate the shell from which the transition originates.*

Examples from the K and L series (in order of increasing energy and decreasing intensity) are:

- K_α line; an electron transition from the L shell to the K shell

- K_β line; an electron transition from the M shell to the K shell

- L_α line; an electron transition from the M shell to the L shell

- L_β line; an electron transition from the N shell to the L shell

28.7 Atomic Transitions and Lasers

Stimulated absorption (upward atomic transition) occurs when photons incident on a gas have energies which exactly match the energy separation between two allowed states of an atom, usually the ground state and a higher energy state. The absorption process raises an atom to various higher energy states called **excited states.**

An atom in an excited state has a certain probability of returning to its original energy state by emission of a photon. This process is called **spontaneous emission.**

When a photon with an energy equal to the excitation energy of an excited atom is incident on the atom, it can increase the probability of de-excitation. This is called **stimulated emission (downward atomic transition)** and results in a second photon. *The incident and the emitted photons are identical; they have equal energies, and they are exactly in phase.*

Population inversion in a gas sample is the condition in which there are more atoms in the excited state than there are in the ground state.

Lasers are monochromatic, coherent light sources that work on the principle of stimulated emission of radiation by a system of atoms.

The following conditions must be satisfied in order to achieve laser action:

- **Population inversion.** There must be a greater number of atoms in an excited state than in the ground state.

- **Metastable states.** The lifetime of the excited states must be long compared with the usually short lifetimes of excited states. Under these conditions stimulated emission will occur before spontaneous emission.

- **Photon confinement.** The emitted photons must remain in the system long enough to allow them to stimulate further emission from other excited atoms. This is achieved by the use of reflecting mirrors at the ends of the system. One end is made totally reflecting and the other is slightly transparent to allow the laser beam to escape.

EQUATIONS AND CONCEPTS

Wavelengths in the Balmer series (the four visible lines) in the emission spectrum of hydrogen can be calculated using empirical Equation (28.1). R_H is a constant called the **Rydberg constant.** The short wavelength limit of the Balmer series is 364.6 nm.

$$R_H = \frac{k_e e^2}{2a_0 hc} = 1.097\,373\,2 \times 10^7\,\text{m}^{-1}$$

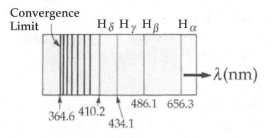

Lines in the Balmer series in the hydrogen spectrum.

$$\frac{1}{\lambda} = R_H \left(\frac{1}{2^2} - \frac{1}{n^2} \right) \quad (n = 3, 4, 5, \ldots) \quad (28.1)$$

An **electron transition** from an initial stationary state, E_i, to a lower energy state, E_f, results in the emission of a photon with a frequency that is proportional to the difference in energies of the initial and final states. *An electron does not radiate energy while in one of the allowed orbits (stationary states) determined by quantization of the orbital angular momentum.*

$$E_i - E_f = hf \qquad (28.4)$$

The angular momentum of the electron about the nucleus must be quantized, always equal to an integral multiple of $n\hbar$. *This condition determines the radii of the allowed electron orbits.*

$$m_e vr = n\hbar \quad (n = 1, 2, 3, \ldots) \qquad (28.5)$$

$$\hbar = \frac{h}{2\pi} = 1.05 \times 10^{-34}\,\text{J} \cdot \text{s}$$

The total energy of the hydrogen atom (*KE* plus *PE* of the proton-electron bound system) depends on the radius of the allowed orbit of the electron. *Note that the total energy of the atom is negative for all values of r except $r = \infty$ when $E = 0$.*

$$E = -\frac{k_e e^2}{2r} \qquad (28.9)$$

The **first three orbits** predicted by Bohr are shown (not to scale) in the figure at right.

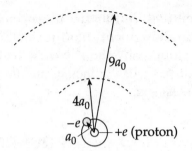

Radii of the allowed orbits have discrete (quantized) values that can be expressed in terms of the Bohr radius (a_0).

$$r_n = \frac{n^2 \hbar^2}{m_e k_e e^2} \quad (n = 1, 2, 3, \ldots) \quad (28.10)$$

The **Bohr radius** (a_0) corresponds to $r = 1$ in Equation (28.10).

$$a_0 = \frac{\hbar^2}{m_e k_e e^2} = 0.052\ 9 \text{ nm} \quad (28.11)$$

$$r_n = n^2 a_0 = n^2 (0.052\ 9 \text{ nm}) \quad (28.12)$$

Quantized energy level values can be expressed in units of electron volts (eV). The lowest allowed energy state or ground state corresponds to the principal quantum number $n = 1$. *The absolute value of the ground state energy is equal to the ionization energy of the atom. The energy level approaches E=0 as r approaches infinity.*

$$E_n = -\frac{13.6}{n^2} \text{ eV} \quad (28.14)$$

A **photon of frequency** f (and wavelength λ) is emitted when an electron undergoes a transition from an initial energy level E_i (outer orbit) to a lower level E_f (inner orbit).

$$f = \frac{E_i - E_f}{h} = \frac{m_e k_e^2 e^4}{4\pi \hbar^3} \left(\frac{1}{n_f^2} - \frac{1}{n_i^2} \right) \quad (28.15)$$

$$\text{where } n_f < n_i$$

A **theoretical value for the Rydberg constant** can be calculated by substituting known values into Equation (28.17).

$$R = \frac{m_e k_e^2 e^4}{4\pi c \hbar^3} \quad (28.17)$$

Photon wavelengths can be calculated using Equation (28.16). *Wavelength, rather than frequency, is the usual experimentally measured quantity.*

$$\frac{1}{\lambda} = R \left(\frac{1}{n_f^2} - \frac{1}{n_i^2} \right) \quad (28.16)$$

The **spectral lines** in the hydrogen spectrum can be arranged into several series. Each series corresponds to an assigned value of the principal quantum number of the final energy state (n_f).

Lyman series: $n_f = 1$; $n_i = 2, 3, 4, \ldots$
Balmer series: $n_f = 2$; $n_i = 3, 4, 5, \ldots$
Paschen series: $n_f = 3$; $n_i = 4, 5, 6, \ldots$
Brackett series: $n_f = 4$; $n_i = 5, 6, 7, \ldots$

The energy level diagram for hydrogen shows transitions for the Lyman, Balmer, and Paschen series. Within each series, the energy difference between adjacent energy levels becomes smaller as n_i increases. The **correspondence principle** states that quantum physics is in agreement with classical physics for large values of the principal quantum number (i.e., when the energy difference between quantized levels are small).

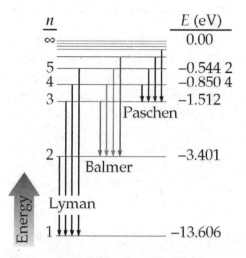

Energy Level Diagram for Hydrogen

Quantum numbers for the hydrogen atom are listed below. *The Pauli exclusion principle states that no two electrons in an atom can have the same set of quantum number values.*

Quantum Number	Name	Allowed Values	Allowed States
n	Principal quantum number	$1, 2, 3, \ldots$	any integer
ℓ	Orbital quantum number	$0, 1, 2, \ldots, (n-1)$	n
m_ℓ	Orbital magnetic quantum number	$-\ell, -\ell+1, -\ell+2, \ldots, 0, 1.2 \ldots \ell$	$2\ell + 1$
m_s	Spin magnetic quantum number	$\tfrac{1}{2}, -\tfrac{1}{2}$	2

REVIEW CHECKLIST

- State the basic postulates of the Bohr model of the hydrogen atom. (Section 28.3)

- Sketch an energy level diagram for hydrogen (assign values of the principal quantum number, n), show transitions corresponding to spectral lines in the several known series, and make calculations of wavelength values. (Section 28.3)

- For each of the quantum numbers, n (the principal quantum number), ℓ (the orbital quantum number), m_ℓ (the orbital magnetic quantum number), and m_s (the spin magnetic quantum number): (i) qualitatively describe what each implies concerning atomic structure, (ii) state the allowed values which may be assigned to each, and (iii) give the number of allowed states which may exist in a particular atom corresponding to each allowed value of the principal quantum number. (Sections 28.4 and 28.5)

- State the Pauli exclusion principle and describe its relevance to the periodic table of the elements. Show how the exclusion principle leads to the known electronic ground state configuration of the light elements. (Section 28.5)

SOLUTIONS TO SELECTED END-OF-CHAPTER PROBLEMS

3. The "size" of the *atom* in Rutherford's model is about 1.0×10^{-10} m. (a) Determine the attractive electrostatic force between an electron and a proton separated by this distance. (b) Determine (in eV) the electrostatic potential energy of the atom.

Solution

(a) From Coulomb's law, the magnitude of the electrical force between two particles having charges q_1 and q_2, when separated by distance r, is

$$F = \frac{k_e |q_1||q_2|}{r^2}$$

Thus, if an electron and proton are separated by a distance of 1.0×10^{-10} m, the magnitude of the attractive force between these unlike charges is

$$F = \frac{\left(8.99 \times 10^9 \ \text{N} \cdot \text{m}^2/\text{C}^2\right)\left(1.60 \times 10^{-19} \ \text{C}\right)^2}{\left(1.0 \times 10^{-10} \ \text{m}\right)^2} = 2.3 \times 10^{-8} \ \text{N}$$ ◊

(b) The electrostatic potential energy of an atom consisting of an electron and a proton separated by distance $r = 1.0 \times 10^{-10}$ m is given by

$$PE = \frac{k_e q_1 q_2}{r} = \frac{\left(8.99 \times 10^9 \ \text{N} \cdot \text{m}^2/\text{C}^2\right)\left(-1.60 \times 10^{-19} \ \text{C}\right)\left(+1.60 \times 10^{-19} \ \text{C}\right)}{1.0 \times 10^{-10} \ \text{m}}$$

or

$$PE = -2.3 \times 10^{-18} \ \text{J}\left(\frac{1 \ \text{eV}}{1.60 \times 10^{-19} \ \text{J}}\right) = -14 \ \text{eV}$$ ◊

7.   A hydrogen atom is in its first excited state ($n = 2$). Using the Bohr theory of the atom, calculate (a) the radius of the orbit, (b) the linear momentum of the electron, (c) the angular momentum of the electron, (d) the kinetic energy, (e) the potential energy, and (f) the total energy.

Solution

(a)   In the Bohr theory, the radii of the allowed orbits are $r_n = n^2 a_0$ where $a_0 = 0.052\,9$ nm. Thus, in the $n = 2$ state, the radius of the orbit is

$$r_2 = (2)^2 (0.052\,9 \text{ nm}) = 0.212 \text{ nm}$$
◊

(b)   The electrical attraction between the electron and proton supplies the centripetal acceleration of the electron. Therefore,

$$m_e \frac{v_n^2}{r_n} = \frac{k_e e^2}{r_n^2} \quad \text{which yields} \quad v_n = \sqrt{\frac{k_e e^2}{m_e r_n}} \quad \text{and} \quad p_n = m_e v_n = \sqrt{\frac{m_e k_e e^2}{r_n}}$$

The linear momentum of the electron in the $n = 2$ state is then

$$p_2 = \sqrt{\frac{m_e k_e e^2}{r_2}} = \sqrt{\frac{(9.11 \times 10^{-31} \text{ kg})(8.99 \times 10^9 \text{ N} \cdot \text{m}^2 / \text{C}^2)(1.60 \times 10^{-19} \text{ C})^2}{0.212 \times 10^{-9} \text{ m}}}$$

or $p_2 = 9.95 \times 10^{-25} \text{ kg} \cdot \text{m/s}$
◊

(c)   According to Bohr's postulates, the angular momentum of the electron is quantized, having values given by $L_n = n(h/2\pi)$. Thus, in the $n = 2$ state, we have

$$L_2 = 2\left(\frac{h}{2\pi}\right) = 2\left(\frac{6.63 \times 10^{-34} \text{ J} \cdot \text{s}}{2\pi}\right) = 2.11 \times 10^{-34} \text{ J} \cdot \text{s}$$
◊

(d)   The kinetic energy of the orbiting electron is $KE_n = \frac{1}{2} m_e v_n^2 = p_n^2 / 2m_e$

When $n = 2$, we use the result from part (b) and find

$$KE_2 = \frac{1}{2} m_e v_2^2 = \frac{p_2^2}{2m_e} = \frac{(9.95 \times 10^{-25} \text{ kg} \cdot \text{m/s})^2}{2(9.11 \times 10^{-31} \text{ kg})}$$

or $KE_2 = 5.44 \times 10^{-19} \text{ J} \left(\dfrac{1 \text{ eV}}{1.60 \times 10^{-19} \text{ J}}\right) = 3.40 \text{ eV}$
◊

(e) The electrical potential energy between two point charges separated by distance r is $PE = k_e q_1 q_2 / r$. Thus, for a hydrogen atom in the $n = 2$ state,

$$PE_2 = \frac{k_e(-e)(+e)}{r_2} = -\frac{(8.99 \times 10^9 \text{ N} \cdot \text{m}^2 / \text{C}^2)(1.60 \times 10^{-19} \text{ C})^2}{0.212 \times 10^{-9} \text{ m}}$$

or $\quad PE_2 = -1.09 \times 10^{-18} \text{ J} \left(\dfrac{1 \text{ eV}}{1.60 \times 10^{-19} \text{ J}} \right) = -6.80 \text{ eV}$ ◊

(f) The total energy of the hydrogen when the electron is in the $n = 2$ state is

$$E_2 = KE_2 + PE_2 = +3.40 \text{ eV} - 6.80 \text{ eV} = -3.40 \text{ eV}$$ ◊

---

11. A hydrogen atom emits a photon of wavelength 656 nm. From what energy orbit to what lower energy orbit did the electron jump?

Solution

The energy levels in the hydrogen atom are given by

$$E_n = -\frac{13.6 \text{ eV}}{n^2} \qquad \text{where} \qquad n = 1, 2, 3, 4, \ldots$$

Thus, the energies of some lower levels are:

$$E_1 = -13.6 \text{ eV}, \; E_2 = -3.40 \text{ eV}, \; E_3 = -1.51 \text{ eV}, \; E_4 = -0.850 \text{ eV}, \ldots$$

When the electron in the atom makes a transition from a level having energy E_i to a level having lower energy E_f, the excess energy is emitted in the form of a photon with

$$E_{\text{photon}} = E_i - E_f$$

If the photon is observed to have wavelength $\lambda = 656$ nm, its energy is

$$E_{\text{photon}} = \frac{hc}{\lambda} = \frac{(6.63 \times 10^{-34} \text{ J} \cdot \text{s})(3.00 \times 10^8 \text{ m/s})}{656 \times 10^{-9} \text{ m}} = 3.03 \times 10^{-19} \text{ J}$$

or

$$E_{\text{photon}} = 3.03 \times 10^{-19} \text{ J} \left(\frac{1 \text{ eV}}{1.60 \times 10^{-19} \text{ J}} \right) = 1.89 \text{ eV}$$

Observe that this energy equals the difference in the energies of the $n = 3$ and $n = 2$ levels in the atom, or $E_{\text{photon}} = E_3 - E_2$. Hence, the transition made by the electron was from the $n = 3$ level to the $n = 2$ level. ◊

---

14. A hydrogen atom initially in its ground state ($n = 1$) absorbs a photon and ends up in the state for which $n = 3$. (a) What is the energy of the absorbed photon? (b) If the atom eventually returns to the ground state, what photon energies could the atom emit?

Solution

(a) The energy of the hydrogen atom when the electron is in the state associated with the quantum number n is

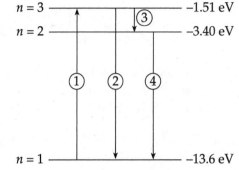

$$E_n = -\frac{13.6 \text{ eV}}{n^2} \quad \text{where} \quad n = 1, 2, 3, \ldots$$

Thus, the energies of the three lowest energy states are $E_1 = -13.6$ eV, $E_2 = -3.40$ eV, and $E_3 = -1.51$ eV as shown in the energy level diagram at the right.

If the electron is to be excited from the ground state ($n = 1$) to the second excited state ($n = 3$) by absorption of a photon as shown by transition 1 in the above diagram, the energy of the absorbed photon must be

$$E_{\text{photon}} = E_3 - E_1 = -1.51 \text{ eV} - (-13.6 \text{ eV}) = 12.1 \text{ eV} \qquad \Diamond$$

(b) As the atom returns to the ground state after the electron is excited to the $n = 3$ state, there are three possible transitions the electron could make as shown by the vertical arrows labeled 2, 3, and 4 in the above diagram. The photon energies produced by each of these transitions are

Transition 2: $\quad E_{\text{photon}} = E_3 - E_1 = -1.51 \text{ eV} - (-13.6 \text{ eV}) = 12.1 \text{ eV} \qquad \Diamond$

Transition 3: $\quad E_{\text{photon}} = E_3 - E_2 = -1.51 \text{ eV} - (-3.40 \text{ eV}) = 1.89 \text{ eV} \qquad \Diamond$

Transition 4: $\quad E_{\text{photon}} = E_2 - E_1 = -3.40 \text{ eV} - (-13.6 \text{ eV}) = 10.2 \text{ eV} \qquad \Diamond$

---

17. (a) If an electron makes a transition from the $n = 4$ Bohr orbit to the $n = 2$ orbit, determine the wavelength of the photon created in the process. (b) Assuming that the atom was initially at rest, determine the recoil speed of the hydrogen atom when this photon is emitted.

Solution

(a) When the electron in a hydrogen atom makes a transition from an orbit with principal quantum number n_i to an orbit with quantum number n_f, the energy of the emitted photon is

$$E_{\text{photon}} = E_{n_i} - E_{n_f} = -\frac{13.6 \text{ eV}}{n_i^2} - \left(-\frac{13.6 \text{ eV}}{n_f^2}\right) = (13.6 \text{ eV})\left(\frac{1}{n_f^2} - \frac{1}{n_i^2}\right)$$

If $n_i = 4$ and $n_f = 2$, the photon energy is

$$E_{photon} = (13.6 \text{ eV})\left(\frac{1}{2^2} - \frac{1}{4^2}\right) = 2.55 \text{ eV}$$

The wavelength of this photon is

$$\lambda = \frac{hc}{E_{photon}} = \frac{(6.63 \times 10^{-34} \text{ J} \cdot \text{s})(3.00 \times 10^8 \text{ m/s})}{2.55 \text{ eV}(1.60 \times 10^{-19} \text{ J/eV})} = 4.88 \times 10^{-7} \text{ m} = 488 \text{ nm} \qquad \lozenge$$

(b) The linear momentum of the system was zero before photon emission and must therefore be zero after this event. Thus, the linear momentum of the recoiling atom and that of the photon must have equal magnitudes but opposite directions.

Equating magnitudes: $\qquad\qquad\qquad P_{atom} = P_{photon}$

or $\qquad\qquad\qquad\qquad\qquad m_{atom}v = h/\lambda$

Since, for a hydrogen atom, $\qquad\quad m_{atom} \approx m_{proton} = 1.67 \times 10^{-27} \text{ kg}$

this gives the recoil speed of the atom as

$$v = \frac{h}{m_{atom}\lambda} = \frac{6.63 \times 10^{-34} \text{ J} \cdot \text{s}}{(1.67 \times 10^{-27} \text{ kg})(4.88 \times 10^{-7} \text{ m})} = 0.814 \text{ m/s} \qquad \lozenge$$

---

21. An electron is in the second excited orbit of hydrogen, corresponding to $n = 3$. Find (a) the radius of the orbit and (b) the wavelength of the electron in this orbit.

Solution

(a) In the Bohr model of the hydrogen atom, the radii of the allowed orbits are given by $r_n = n^2 a_0$, where $a_0 = 0.052\ 9$ nm and $n = 1, 2, 3, \ldots$. The radius of the second excited orbit is then

$$r_3 = 3^2 (0.052\ 9 \text{ nm}) = 9(0.052\ 9 \text{ nm}) = 0.476 \text{ nm} \qquad \lozenge$$

(b) According to de Broglie's interpretation of one of Bohr's original postulates, the allowed orbits are those in which the electron can form a standing wave pattern. That is, they are the orbits for which the circumference is an integer multiple of the de Broglie wavelength for the electron in those orbits, or the orbits for which

$$2\pi r_n = n\lambda$$

Thus, the wavelength of the electron in the second excited orbit, corresponding to $n = 3$, should be

$$\lambda = \frac{2\pi r_n}{n} = \frac{2\pi r_3}{3} = \frac{2\pi(0.476 \text{ nm})}{3} = 0.997 \text{ nm} \qquad \lozenge$$

26. Using the concept of standing waves, de Broglie was able to derive Bohr's stationary orbit postulate. He assumed that a confined electron could exist only in states where its de Broglie waves form standing-wave patterns, as in Figure 28.6. Consider a particle confined in a box of length L to be equivalent to a string of length L and fixed at both ends. Apply de Broglie's concept to show that (a) the linear momentum of this particle is quantized with $p = mv = nh/2L$ and (b) the allowed states correspond to particle energies of $E_n = n^2 E_0$, where $E_0 = h^2/(8mL^2)$.

Solution

(a) In order to form a standing wave in a string fixed at both ends, it is necessary that the length of the string be an integer multiple of a half wavelength,

$$L = n\left(\frac{\lambda}{2}\right) \qquad \text{or} \qquad \lambda = \frac{2L}{n}$$

According to de Broglie's hypothesis, the wavelength of a particle of mass m is given by

$$\lambda = \frac{h}{p} = \frac{h}{mv} \qquad \text{and} \qquad p = mv = \frac{h}{\lambda}$$

Thus, if we consider the particle confined in a box of length L to be equivalent to a string of length L fixed at both ends, the requirement to form standing waves would be

$$p_n = mv_n = \frac{h}{\lambda} = \frac{h}{2L/n} = \frac{nh}{2L} \qquad \text{where} \qquad n = 1, 2, 3, \dots \qquad \Diamond$$

(b) Considering the particle to be nonrelativistic, the total energy of a particle in a box would be

$$E_n = \frac{1}{2}mv_n^2 = \frac{p_n^2}{2m} = \frac{(nh/2L)^2}{2m} = \frac{n^2 h^2}{8mL^2}$$

$$\text{or} \qquad E_n = n^2 E_0 \qquad \text{where} \qquad E_0 = \frac{h^2}{8mL^2} \qquad \Diamond$$

---

31. Two electrons in the same atom have $n = 3$ and $\ell = 1$. (a) List the quantum numbers for the possible states of the atom. (b) How many states would be possible if the exclusion principle did not apply to the atom?

Solution

(a) The Pauli exclusion principle states that no two electrons in the same atom can have identical sets of quantum numbers (n, ℓ, m_ℓ, and m_s). For states having given values of n and ℓ, the orbital magnetic quantum number, m_ℓ, may vary from $-\ell$ to $+\ell$ in integer steps. For each of these $(2\ell + 1)$ possible values of m_ℓ, the spin magnetic quantum number m_s may take on either of two values, $\pm 1/2$. When $n = 3$ and $\ell = 1$, there are 6 possible states the first electron could be in. For each of these states, there are 5 other states the second electron could be in without violating the exclusion principle, giving a total of $6 \times 5 = 30$ possible combinations of quantum numbers for the two electrons. These combinations are listed below. ◊

For both electrons, $n = 3$ and $\ell = 1$. The other quantum numbers are:

	m_ℓ	m_s	m_ℓ	m_s	m_ℓ	m_s	m_ℓ	m_s	m_ℓ	m_s	m_ℓ	m_s
Electron #1	+1	$+\frac{1}{2}$	+1	$+\frac{1}{2}$	+1	$+\frac{1}{2}$	+1	$-\frac{1}{2}$	+1	$-\frac{1}{2}$	+1	$-\frac{1}{2}$
Electron #2	+1	$-\frac{1}{2}$	0	$\pm\frac{1}{2}$	−1	$\pm\frac{1}{2}$	+1	$+\frac{1}{2}$	0	$\pm\frac{1}{2}$	−1	$\pm\frac{1}{2}$

	m_ℓ	m_s	m_ℓ	m_s	m_ℓ	m_s	m_ℓ	m_s	m_ℓ	m_s	m_ℓ	m_s
Electron #1	0	$+\frac{1}{2}$	0	$+\frac{1}{2}$	0	$+\frac{1}{2}$	0	$-\frac{1}{2}$	0	$-\frac{1}{2}$	0	$-\frac{1}{2}$
Electron #2	+1	$\pm\frac{1}{2}$	0	$-\frac{1}{2}$	−1	$\pm\frac{1}{2}$	+1	$\pm\frac{1}{2}$	0	$+\frac{1}{2}$	−1	$\pm\frac{1}{2}$

	m_ℓ	m_s	m_ℓ	m_s	m_ℓ	m_s	m_ℓ	m_s	m_ℓ	m_s	m_ℓ	m_s
Electron #1	−1	$+\frac{1}{2}$	−1	$+\frac{1}{2}$	−1	$+\frac{1}{2}$	−1	$-\frac{1}{2}$	−1	$-\frac{1}{2}$	−1	$-\frac{1}{2}$
Electron #2	+1	$\pm\frac{1}{2}$	0	$\pm\frac{1}{2}$	−1	$-\frac{1}{2}$	+1	$\pm\frac{1}{2}$	0	$\pm\frac{1}{2}$	−1	$+\frac{1}{2}$

(b) If the electrons did not obey the exclusion principle, there would be 36 possible combinations of quantum numbers for these two electrons. For each of the 6 possible states described above for the first electron, the second electron would have the choice of any of the 6 states (since it would be allowed to duplicate the state of the first electron). This would give a total of $6 \times 6 = 36$ possible combinations. ◊

---

37. The K series of the discrete spectrum of tungsten contains wavelengths of 0.018 5 nm, 0.020 9 nm, and 0.021 5 nm. The K-shell ionization energy is 69.5 keV. Determine the ionization energies of the L, M, and N shells.

Solution

Characteristic x-rays of the K series are produced when an electron in an upper energy level of a many-electron atom drops down to fill a vacancy in the K ($n = 1$) shell. Assuming that the given wavelengths are the three longest wavelengths in the K series of tungsten, the transitions that produce them are from the N ($n = 4$) shell to the K shell, from the M ($n = 3$) shell to the K shell, and from the L ($n = 2$) shell to the K shell, respectively. These transitions are shown in the

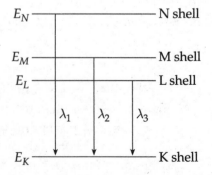

energy level diagram where $\lambda_1 = 0.018\,5$ nm, $\lambda_2 = 0.020\,9$ nm, and $\lambda_3 = 0.021\,5$ nm. If the ionization energy of the K shell is 69.5 keV, then $E_K = -69.5$ keV.

The photon produced in a transition has energy $\qquad E_{photon} = \dfrac{hc}{\lambda} = E_i - E_f$

so the energy of the initial shell is $\qquad\qquad E_i = E_f + \dfrac{hc}{\lambda}$

Thus, $E_N = E_K + \dfrac{hc}{\lambda_1} = -69.5$ keV $+ \dfrac{(6.63 \times 10^{-34}\ \text{J}\cdot\text{s})(3.00 \times 10^8\ \text{m/s})}{0.018\,5 \times 10^{-9}\ \text{m}} \left(\dfrac{1\ \text{keV}}{1.60 \times 10^{-16}\ \text{J}} \right)$

or $\qquad E_N = -2.30$ keV

Similarly, $\qquad E_M = E_K + \dfrac{hc}{\lambda_2} = -10.0$ keV and $E_L = E_K + \dfrac{hc}{\lambda_3} = -11.7$ keV

Thus, the ionization energies of the L, M, and N shells are:

$\qquad$ 11.7 keV, 10.0 keV, and 2.30 keV, respectively $\qquad\qquad\qquad\qquad\qquad\Diamond$

---

41. An electron in chromium moves from the $n = 2$ state to the $n = 1$ state without emitting a photon. Instead, the excess energy is transferred to an outer electron (one in the $n = 4$ state), which is then ejected by the atom. In this Auger (pronounced "ohjay") process, the ejected electron is referred to as an Auger electron. (a) Find the change in energy associated with the transition from $n = 2$ into the vacant $n = 1$ state using Bohr theory. Assume that only one electron in the K shell is shielding part of the nuclear charge. (b) Find the energy needed to ionize an $n = 4$ electron, assuming 22 electrons shield the nucleus. (c) Find the kinetic energy of the ejected (Auger) electron. (All answers should be in electron volts.)

Solution

(a) A filled K shell would contain 2 electrons. With one vacancy in the K shell of chromium ($Z = 24$), an electron in the L shell has one electron shielding it from the nuclear charge, so $Z_{eff} = Z - 1 = 24 - 1 = 23$. The estimated energy the atom gives up during a transition from the L shell to the K shell vacancy is

$$\Delta E = E_i - E_f \approx -\dfrac{Z_{eff}^2 (13.6\ \text{eV})}{n_i^2} - \left[-\dfrac{Z_{eff}^2 (13.6\ \text{eV})}{n_f^2} \right] = Z_{eff}^2 (13.6\ \text{eV}) \left[\dfrac{1}{n_f^2} - \dfrac{1}{n_i^2} \right]$$

$\qquad$ or

$$\Delta E \approx (23)^2 (13.6\ \text{eV}) \left[\dfrac{1}{1^2} - \dfrac{1}{2^2} \right] = 5.40 \times 10^3\ \text{eV} = 5.40\ \text{keV} \qquad\qquad\qquad \Diamond$$

(b) With a vacancy in the K shell, we assume that $Z - 2 = 24 - 2 = 22$ electrons shield the outermost electron (in a $4s$ state) from the nuclear charge. Thus, for this outer electron, $Z_{eff} = 24 - 22 = 2$ and the estimated energy required to remove this electron from the atom is

$$E_{ionization} = E_f - E_i = 0 - E_i \approx -\left[-\frac{Z_{eff}^2 (13.6 \text{ ev})}{n_i^2} \right] = \frac{2^2 (13.6 \text{ ev})}{4^2} = 3.40 \text{ eV} \qquad \Diamond$$

(c) The atom transfers energy ΔE to this outer electron, which spends an amount $E_{ionization}$ of it in escaping from the atom. The remaining energy appears as kinetic energy of the now free electron. Therefore,

$$KE = \Delta E - E_{ionization} = 5.40 \times 10^3 \text{ eV} - 3.40 \text{ eV} \approx 5.40 \times 10^3 \text{ eV} = 5.40 \text{ keV} \qquad \Diamond$$

---

45. Use Bohr's model of the hydrogen atom to show that, when the atom makes a transition from the state n to the state $n - 1$, the frequency of the emitted light is given by

$$f = \frac{2\pi^2 m_e k_e^2 e^4}{h^3} \left(\frac{2n - 1}{(n-1)^2 n^2} \right)$$

Solution

According to the Bohr model of the hydrogen atom, the frequency of the photon emitted when the electron makes a transition from state n to state $n - 1$ is

$$f = \frac{\Delta E}{h} = \frac{E_n - E_{n-1}}{h}$$

From Equation 28.13 in the textbook, the energy levels in hydrogen are given in terms of fundamental constants as

$$E_n = -\frac{m_e k_e^2 e^4}{2\hbar^2 n^2} = -\frac{2\pi^2 m_e k_e^2 e^4}{h^2 n^2} \qquad \text{where} \qquad n = 1, 2, 3, \ldots$$

Thus, the photon emitted in a transition from state n to state $n - 1$ has frequency

$$f = \frac{1}{h} \left[-\frac{2\pi^2 m_e k_e^2 e^4}{h^2 n^2} - \left(-\frac{2\pi^2 m_e k_e^2 e^4}{h^2 (n-1)^2} \right) \right] = \frac{2\pi^2 m_e k_e^2 e^4}{h^3} \left[\frac{1}{(n-1)^2} - \frac{1}{n^2} \right]$$

or

$$f = \frac{2\pi^2 m_e k_e^2 e^4}{h^3} \left[\frac{n^2 - (n-1)^2}{(n-1)^2 n^2} \right] = \frac{2\pi^2 m_e k_e^2 e^4}{h^3} \left[\frac{n^2 - (n^2 - 2n + 1)}{(n-1)^2 n^2} \right]$$

which reduces to

$$f = \frac{2\pi^2 m_e k_e^2 e^4}{h^3} \left[\frac{2n - 1}{(n-1)^2 n^2} \right] \qquad \Diamond$$

29

Nuclear Physics

NOTES FROM SELECTED CHAPTER SECTIONS

29.1 Some Properties of Nuclei

Important terms in the description of nuclear properties include:

- The **atomic number, Z**, equals the number of protons in the nucleus.

- The **neutron number, N**, equals the number of neutrons in the nucleus.

- The **mass number, A**, equals the number of nucleons (neutrons plus protons) in the nucleus.

Nuclei can be represented symbolically as $_Z^A X$, where X designates the chemical symbol for a specific chemical element.

The nuclei of all atoms of a particular element contain the same number of protons but may contain different numbers of neutrons. Nuclei that are related in this way are called **isotopes. The isotopes of an element have the same Z value but different N and A values.**

The **unified mass unit, u, is defined such that the mass of the isotope ^{12}C is exactly 12 u** ($1 \, u = 1.660\,559 \times 10^{-27} \, kg$).

Based on nuclear scattering and other experiments, it is possible to conclude that:

- Most nuclei are approximately spherical, and all have nearly the same density.

- Nucleons combine to form a nucleus as though they were tightly packed spheres.

- The stability of nuclei is due to a short-range attractive nuclear force.

- The nuclear force is approximately the same for an interaction between any pair of nucleons (p-p, n-n, or p-n).

29.2 Binding Energy

The total mass of a nucleus is always less than the sum of the masses of its individual nucleons. This difference in mass is the origin of the nuclear **binding energy** and represents the energy that would be required to separate the nucleus into neutrons and protons.

29.3 Radioactivity

The decay of radioactive substances can be accompanied by the emission of three forms of radiation that vary in ability to penetrate shielding materials:

- **alpha** (α) **particles** — Helium nuclei (^4_2He), which can barely penetrate a sheet of paper.

- **beta** (β) **particles** — Either electrons (e^-) or positrons (e^+) able to penetrate a few millimeters of aluminum.

- **gamma** (γ) **rays** — High-energy photons capable of penetrating several centimeters of lead.

A positron is the antiparticle of the electron; it has a mass equal to m_e and a charge of $+q_e$.

The decay rate, or activity, R, of a sample of a radioactive substance is defined as the number of decays occurring per second. The half-life $(T_{1/2})$ of a radioactive substance is the time during which half of an initial number of radioactive nuclei in a sample will decay. The customary unit of radioactivity is the Curie (Ci); the SI unit of activity is the **becquerel** (Bq), where 1 Bq = 1 decay/s and 1 Ci = 3.7×10^{10} Bq.

29.4 The Decay Processes

The overall decay process can be represented in equation form as:

$$\underset{\substack{\text{Parent} \\ \text{nucleus}}}{\text{X}} \rightarrow \underset{\substack{\text{Daughter} \\ \text{nucleus}}}{\text{Y}} + \underset{\substack{\text{Emitted} \\ \text{radiation}}}{}$$

In an equation representing a specific radioactive decay process, the sum of the mass numbers on each side of the equation must be equal, and the sum of the atomic numbers on each side of the equation must be equal.

The neutrino and antineutrino shown in the beta decay processes in the table below are required for conservation of energy and momentum.

The neutrino has the following properties:

- zero electric charge

- rest mass much smaller than that of the electron (recent experiments suggest that the neutrino mass may not be zero)

- spin of $\frac{1}{2}$ (satisfying the law of conservation of angular momentum)

- very weak interaction with matter (therefore very difficult to detect)

General characteristics of alpha, beta, and gamma decay are summarized below.

Decay	can be written as	Process
Alpha decay	$_Z^A X \rightarrow \, _{Z-2}^{A-4} Y + \, _2^4 He$	The parent nucleus emits an alpha particle and loses two protons and two neutrons.
Beta decay	$_Z^A X \rightarrow \, _{Z+1}^A Y + e^- + \bar{\nu}$ (electron emission) $_Z^A X \rightarrow \, _{Z-1}^A Y + e^+ + \nu$ (positron emission)	A nucleus can undergo beta decay in two ways. The parent nucleus can emit an electron (e^-) and an antineutrino ($\bar{\nu}$) or emit a positron (e^+) and a neutrino (ν).
Gamma decay	$_Z^A X^* \rightarrow \, _Z^A X + \gamma$ * parent nucleus in an excited state	A nucleus in an excited state decays to a lower energy state (often the ground state) and emits a gamma ray.

29.5   Natural Radioactivity

Natural radioactivity is the decay of radioactive nuclei found in nature.

Artificial radioactivity is the decay of nuclei produced in the laboratory through nuclear reactions.

Most naturally decaying radionuclides are members of a **decay series.**

The Uranium Series, Actinium Series, and Thorium Series each begin with a long half-life, naturally occurring radioisotope and, following a sequence of decays, ends in a stable isotope. A fourth series, the Neptunium Series, begins with the decay of an artificially produced radionuclide.

29.6   Nuclear Reactions

Nuclear reactions are events in which the structure or properties of colliding nuclei are altered. The ***Q*-value** is the energy required to balance the overall reaction equation.

Exothermic reactions release energy and have positive *Q*-values. The *Q*-values of endothermic reactions are negative, and the minimum energy of the incoming particle for which an endothermic reaction will occur is called the **threshold energy.**

29.7   Medical Applications of Radiation

Radiation damage to cells in biological organisms is primarily due to ionizing events and can be separated into two categories:

- **Somatic damage** is radiation damage to cells other than reproductive cells.

- **Genetic damage** affects only reproductive cells.

Several units are used to quantify radiation exposure and dose:

- **Roengten (R);** the amount of ionizing radiation that will produce 2.08×10^9 ion pairs in 1 cm³ of air under standard conditions.

- **rad;** the amount of radiation that deposits 10^{-2} J of energy into 1 kg of absorbing material.

- **RBE** (relative biological effectiveness) factor; the number of rad of either x-ray or gamma radiation that will produce the same degree of biological damage as the actual radiation being used (e.g., electrons, protons, neutrons, heavy ions, etc.).

- **rem** (roentgen equivalent in man); the product of the dose measured in rads and the RBE factor. One rem of any two types of radiation will produce the same degree of biological damage. Dose in rem = dose in rad × RBE.

Current SI units of radiation dose and exposure are the gray (Gy) and the sievert (Sv). 1 Sv = 100 rem and 1 Gy = 100 rad.

EQUATIONS AND CONCEPTS

The **average nuclear radius** is proportional to the cube root of the mass number (or total number of nucleons). This means that the volume is proportional to A, nuclei are approximately spherical in shape, and all nuclei have nearly the same density.

$$r = r_0 A^{1/3} \tag{29.1}$$

$$r_0 = 1.2 \text{ fm}$$

The **fermi (fm)** is a convenient unit of length to use in stating nuclear dimensions.

$$1 \text{ fm} = 10^{-15} \text{ m}$$

The **decay constant, λ,** is characteristic of a particular isotope. The number of radioactive nuclei in a given sample which undergo decay during a time interval Δt is proportional to the number of nuclei present.

$$\Delta N = -\lambda N \Delta t \tag{29.2}$$

The decay rate or activity, R, of a sample of radioactive nuclei is defined as the number of decays per second. Activity, R, can be expressed in units of becquerels or curies.

$$R = \left| \frac{\Delta N}{\Delta t} \right| = \lambda N \qquad (29.3)$$

$$1 \text{ Ci} \equiv 3.7 \times 10^{10} \text{ decays/s} \qquad (29.6)$$

$$1 \text{ Bq} = 1 \text{ decay/s} \qquad (29.7)$$

The **decay curve** is a plot of N vs. t (the number of nuclei remaining in a sample versus the elapsed time). *The number of nuclei remaining in a sample of a radioactive substance decreases exponentially with time.* N_0 is the number of nuclei in the sample at $t = 0$, and time is shown in units of half-lives.

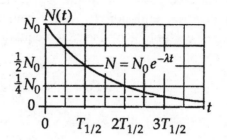

Equation (29.4b) gives the number of remaining nuclei after an elapsed time of n half-lives.

$$N = N_0 e^{-\lambda t} \qquad (29.4a)$$

$$N = N_0 \left(\frac{1}{2} \right)^n \qquad (29.4b)$$

The **half-life of a radioactive substance** is the time required for half of the radioactive nuclei remaining in a sample of the substance to decay.

$$T_{1/2} = \frac{\ln 2}{\lambda} = \frac{0.693}{\lambda} \qquad (29.5)$$

Spontaneous decay of radioactive nuclei proceeds by one the following processes: Alpha decay, beta decay, electron capture, or gamma decay. The overall decay process can be represented by a decay equation.

$$X \quad \rightarrow \quad Y \quad + \quad \text{Emitted radiation}$$

Parent nucleus   Daughter nucleus

Alpha Decay

When a nucleus decays by alpha emission, the parent nucleus loses two neutrons and two protons. For alpha emission to occur, the mass of the parent nucleus $\left({}_{Z}^{A}X \right)$ must be greater than the combined masses of the daughter nucleus $\left({}_{Z-2}^{A-4}Y \right)$ and the emitted alpha particle. The decay of ${}_{92}^{238}U$ is shown in Equation (29.9)

$${}_{Z}^{A}X \rightarrow {}_{Z-2}^{A-4}Y + {}_{2}^{4}He \qquad (29.8)$$

$${}_{92}^{238}U \rightarrow {}_{90}^{234}Th + {}_{2}^{4}He \qquad (29.9)$$

For ${}_{92}^{238}U$, $T_{1/2} = 4.47 \times 10^9$ yr

Beta decay

The beta decay process is accompanied by emission of a third particle that is required for conservation of energy and momentum.

Two modes of beta decay are: (1) Electron emission (e⁻) with antineutrino ($\bar{v}$) and (2) Positron emission (e⁺) with neutrino (v). For each mode of decay, the equations at right show:

(i)   The equation representing the process,

(ii) an example of the decay mode, and

(iii) an equation describing the origin of the electron (or proton).

Gamma Decay

Nuclei which undergo beta or alpha decay are often left in an excited energy state (indicated by the * symbol in Equation (29.18). The nucleus returns to the ground state by emission of one or more photons. *Gamma decay results in no change in mass number or atomic number.*

Nuclear reactions can occur when target nuclei (X) are bombarded with energetic particles (a) resulting in a daughter or product nucleus (Y) and an outgoing particle (b). Equation (29.21) describes the first nuclear reaction produced in the laboratory. *In this case both reactants and both products are stable isotopes. In these reactions the structure, identity, or properties of the target nuclei are changed.*

The Q value of a nuclear reaction is the quantity of energy required to balance the equation representing the reaction. *Q is positive for exothermic reactions and negative for endothermic reactions.*

Electron Emission

(i) $_{Z}^{A}X \rightarrow {}_{Z+1}^{A}Y + e^- + \bar{v}$

(ii) $_{6}^{14}C \rightarrow {}_{7}^{14}N + e^- + \bar{v}$ (29.15)

(iii) $_{1}^{1}n \rightarrow {}_{1}^{1}p + e^- + \bar{v}$

Positron Emission

(i) $_{Z}^{A}X \rightarrow {}_{Z-1}^{A}Y + e^+ + v$

(ii) $_{7}^{12}N \rightarrow {}_{6}^{12}C + e^+ + v$ (29.16)

(iii) $_{1}^{1}p \rightarrow {}_{1}^{1}n + e^+ + v$

$$_{Z}^{A}X^* \rightarrow {}_{Z}^{A}X + \gamma$$

$$_{6}^{12}C^* \rightarrow {}_{6}^{12}C + \gamma \qquad (29.18)$$

$$a + X \rightarrow Y + b$$

$$_{2}^{4}He + {}_{7}^{14}N \rightarrow {}_{8}^{17}O + {}_{1}^{1}H \qquad (29.21)$$

$$Q = KE_{products} - KE_{reactants}$$

$$Q = (m_{reactants} - m_{products})c^2$$

A **threshold energy** (minimum kinetic energy of the incident particle) is characteristic of endothermic reactions. This minimum value of kinetic energy of the incident particle is required in order for energy and momentum to be conserved.

$$KE_{min} = \left(1 + \frac{m}{M}\right)|Q| \qquad (29.24)$$

m = mass of incident particle

M = mass of target particle

SUGGESTIONS, SKILLS, AND STRATEGIES

The rest energy of a particle is given by $E = mc^2$. It is therefore often convenient to express the unified mass unit in terms of its energy equivalent, $1\,u = 1.660\,559 \times 10^{-27}$ kg or $1\,u = 931.50$ MeV/c^2. When masses are expressed in units of u, energy values are then $E = m(931.50 \text{ MeV/u})$.

Equation (29.4a) can be solved for the particular time t after which the number of remaining nuclei will be some specified fraction (N/N_0) of the original number N_0. This can be done by taking the natural log of each side of Equation (29.4a) to find

$$t = \left(\frac{1}{\lambda}\right)\ln\left(\frac{N_0}{N}\right)$$

Then substitute for N the specified fraction of N_0.

REVIEW CHECKLIST

- State and apply to the solution of related problems, the formula which expresses decay rate as a function of decay constant and number of radioactive nuclei. Also apply the exponential formula which expresses the number of radioactive nuclei remaining as a function of elapsed time, decay constant, or half-life, and the initial number of nuclei. (Section 29.3)

- Identify each of the components of radiation that are emitted by nuclei through natural radioactive decay and describe the basic properties of each. Write out typical equations to illustrate the processes of alpha decay, beta decay, and gamma emission. (Section 29.4)

- Calculate the Q value of a given nuclear reaction and determine the threshold energy of an endothermic reaction. (Section 29.6)

- Given the reactants and one of the products of a nuclear reaction, be able to balance the reaction equation and identify the remaining product nuclide. (Section 29.6)

- Be familiar with the various units used to express quantities of radiation dose. (Section 29.7)

SOLUTIONS TO SELECTED END-OF-CHAPTER PROBLEMS

5. An alpha particle ($Z = 2$, mass 6.64×10^{-27} kg) approaches to within 1.00×10^{-14} m of a carbon nucleus ($Z = 6$). What are (a) the maximum Coulomb force on the alpha particle, (b) the acceleration of the alpha particle at this time, and (c) the potential energy of the alpha particle at the same time?

Solution

(a) The repulsive electrical force exerted on the alpha particle by the carbon nucleus has a magnitude of

$$F = \frac{k_e q_1 q_2}{r^2} = \frac{k_e (2e)(6e)}{r^2} = \frac{12 k_e e^2}{r^2}$$

The maximum of this force occurs at the distance of closest approach, so

$$F_{max} = \frac{12 k_e e^2}{r_{min}^2} = \frac{12 (8.99 \times 10^9 \text{ N} \cdot \text{m}^2/\text{C}^2)(1.60 \times 10^{-19} \text{ C})^2}{(1.00 \times 10^{-14} \text{ m})^2} = 27.6 \text{ N} \qquad \lozenge$$

(b) At the point of closest approach, the acceleration of the alpha particle is

$$a_{max} = \frac{F_{max}}{m_\alpha} = \frac{27.6 \text{ N}}{6.64 \times 10^{-27} \text{ kg}} = 4.16 \times 10^{27} \text{ m/s}^2 \qquad \lozenge$$

(c) The electrical potential energy of a point charge q_1 at distance r from a second point charge q_2 is

$$PE_e = q_1 V = q_1 \left(\frac{k_e q_2}{r} \right) = \frac{k_e q_1 q_2}{r}$$

Thus, the potential energy of the alpha particle at the point of its closest approach to the carbon nucleus is

$$PE_{max} = \frac{k_e (2e)(6e)}{r_{min}} = \frac{12 (8.99 \times 10^9 \text{ N} \cdot \text{m}^2/\text{C}^2)(1.60 \times 10^{-19} \text{ C})^2}{1.00 \times 10^{-14} \text{ m}}$$

or $\quad PE_{max} = 2.76 \times 10^{-13} \text{ J} \left(\frac{1 \text{ MeV}}{1.60 \times 10^{-13} \text{ J}} \right) = 1.73 \text{ MeV} \qquad \lozenge$

11. A pair of nuclei for which $Z_1 = N_2$ and $Z_2 = N_1$ are called *mirror isobars*. (The atomic and neutron numbers are interchangeable.) Binding-energy measurements on such pairs can be used to obtain evidence of the charge independence of nuclear forces. Charge independence means that the proton-proton, proton-neutron, and neutron-neutron forces are approximately equal. Calculate the difference in binding energy for the two mirror nuclei $^{15}_{8}O$ and $^{15}_{7}N$.

Solution

The binding energy of a nucleus $^A_Z X$ is given by $E_b = (\Delta m)c^2$, where the mass deficit is

$$\Delta m = \left[Zm_{^1_1 H} + (A - Z)m_n \right] - m_{^A_Z X}$$

Here, Z is the atomic number (number of protons in the nucleus), $(A - Z)$ is the number of neutrons, m_n is the mass of a neutron, $m_{^1_1 H}$ is the mass of a neutral hydrogen atom, and $m_{^A_Z X}$ is the mass of a neutral atom containing the nucleus of interest. We may use atomic masses from Appendix B of the textbook in this calculation since the included masses of Z electrons will cancel in the subtraction process.

For $^{15}_{8}O$, $\qquad \Delta m = 8(1.007\ 825\ u) + 7(1.008\ 665\ u) - 15.003\ 065\ u = 0.120\ 190\ u$

and the binding energy is $E_b = 0.120\ 190\ u \cdot c^2$. The energy equivalent of the atomic mass unit is $1\ u = 931.5\ MeV/c^2$, so the binding energy of $^{15}_{8}O$ is

$$E_b = (0.120\ 190\ u)(931.5\ MeV/u) = 112.0\ MeV$$

For $^{15}_{7}N$, $\qquad \Delta m = 7(1.007\ 825\ u) + 8(1.008\ 665\ u) - 15.000\ 109\ u = 0.123\ 986\ u$

and $\qquad E_b = (0.123\ 986\ u)(931.5\ MeV/u) = 115.5\ MeV$

Thus, the difference in the binding energies of these mirror isobars is

$$\Delta E_b = E_b\big|_{^{15}N} - E_b\big|_{^{15}O} = 115.5\ MeV - 112.0\ MeV = 3.5\ MeV \qquad \qquad \Diamond$$

---

15. The half-life of an isotope of phosphorus is 14 days. If a sample contains 3.0×10^{16} such nuclei, determine its activity. Express your answer in curies.

Solution

The activity, R, of a radioactive sample is directly proportional to the number of radioactive nuclei in that sample. That is,

$$R = \lambda N$$

where N is the number of nuclei in the sample, and λ is the decay constant of that species of unstable nuclei.

If R_0 is the activity of the sample at time $t = 0$, the activity at any later time is

$$R = R_0 e^{-\lambda t}$$

When $t = T_{1/2}$ (the half-life for this type nucleus), the activity is $R = R_0/2$, so

$$\frac{1}{2} R_0 = R_0 e^{-\lambda T_{1/2}} \quad \text{or} \quad e^{\lambda T_{1/2}} = 2$$

This yields $\quad \lambda T_{1/2} = \ln 2 \quad$ or $\quad \lambda = \dfrac{\ln 2}{T_{1/2}}$

Thus, if $T_{1/2} = 14$ d, the decay constant is $\quad \lambda = \dfrac{\ln 2}{14 \text{ d}}$

and the activity in curies ($1 \text{ Ci} = 3.7 \times 10^{10}$ decays per second) of a sample containing $N = 3.0 \times 10^{16}$ radioactive nuclei of this type is

$$R = \lambda N = \left(\frac{\ln 2}{14 \text{ d}}\right)(3.0 \times 10^{16})\left(\frac{1 \text{ d}}{86 \ 400 \text{ s}}\right)\left(\frac{1 \text{ Ci}}{3.7 \times 10^{10} \text{ s}^{-1}}\right) = 0.46 \text{ Ci} \qquad \lozenge$$

---

19. Many smoke detectors use small quantities of the isotope ^{241}Am in their operation. The half-life of ^{241}Am is 432 yr. How long will it take for the activity of this material to decrease to 1.00×10^{-3} of the original activity?

Solution

The activity of a radioactive sample at time t is given by

$$R = R_0 e^{-\lambda t}$$

where R_0 is the original activity (at time $t = 0$) and λ is the decay constant for this type nuclei. The decay constant is related to the half-life by the expression

$$\lambda = \frac{\ln 2}{T_{1/2}}$$

Thus, the decay constant for ^{241}Am is $\quad \lambda = \dfrac{\ln 2}{432 \text{ yr}}$

and when the activity has decreased to 1.00×10^{-3} of the original value,

$$\frac{R_0}{1\,000} = R_0 e^{-\lambda t} \quad \text{or} \quad e^{\lambda t} = 1\,000$$

Taking the natural logarithm of both sides of the last result yields

$$\lambda t = \ln(1\,000)$$

or $\quad t = \dfrac{\ln(1\,000)}{\lambda} = \dfrac{\ln(1\,000)}{\ln 2/432 \text{ yr}} = (432 \text{ yr})\dfrac{\ln(1\,000)}{\ln 2} = 4.31 \times 10^3 \text{ yr}$ $\qquad \lozenge$

---

21. A freshly prepared sample of a certain radioactive isotope has an activity of 10.0 mCi. After 4.00 h, the activity is 8.00 mCi. (a) Find the decay constant and half-life of the isotope. (b) How many atoms of the isotope were contained in the freshly prepared sample? (c) What is the sample's activity 30 h after it is prepared?

Solution

(a) The activity of a radioactive sample varies with time according to $R = R_0 e^{-\lambda t}$, where R_0 is the activity at time $t = 0$ and λ is the decay constant for that species of radioactive nuclei. Thus, if $R = R_0 = 10.0$ mCi at $t = 0$ and the activity at $t = 4.0$ h is $R = 8.00$ mCi, we have $e^{-\lambda(4.00 \text{ h})} = R/R_0 = 0.800$. Taking the natural logarithm of both sides of this result gives $-\lambda(4.00 \text{ h}) = \ln(0.800)$, and the decay constant for this isotope is

$$\lambda = \frac{\ln 0.800}{-4.00 \text{ h}} = 5.58 \times 10^{-2} \text{ h}^{-1} \qquad \lozenge$$

The half-life is related to the decay constant by the expression $T_{1/2} = \ln 2/\lambda$. The half-life of this isotope is then

$$T_{1/2} = \frac{\ln 2}{5.58 \times 10^{-2} \text{ h}^{-1}} = 12.4 \text{ h} \qquad \lozenge$$

(b) The activity of a radioactive sample is related to the number of radioactive nuclei in the sample by $R = \lambda N$, so the number of atoms of this isotope in the freshly made sample was

$$N_0 = \frac{R_0}{\lambda} = \frac{10.0 \times 10^{-3} \text{ Ci}}{5.58 \times 10^{-2} \text{ h}^{-1}} \left(\frac{3\,600 \text{ s}}{1 \text{ h}} \right) \left(\frac{3.70 \times 10^{10} \text{ s}^{-1}}{1 \text{ Ci}} \right) = 2.39 \times 10^{13} \qquad \lozenge$$

(c) After an elapsed time of $t = 30$ h, the activity of this sample will be

$$R = R_0 e^{-\lambda t} = (10.0 \text{ mCi}) e^{-(5.58 \times 10^{-2} \text{ h}^{-1})(30 \text{ h})} = 1.9 \text{ mCi} \qquad \lozenge$$

---

27. Determine which of the following suggested decays can occur spontaneously:

(a) $^{40}_{20}\text{Ca} \rightarrow \text{e}^+ + ^{40}_{19}\text{K}$
(b) $^{144}_{60}\text{Nd} \rightarrow ^4_2\text{He} + ^{140}_{58}\text{Ce}$

Solution

For a decay to occur spontaneously, a release of energy (as indicated by a positive Q-value) must occur during the decay. The Q-value is given by $Q = (\Delta m)c^2$, where Δm is the difference between the mass of the original nucleus and the total mass of the decay products.

(a) When the decay products include positrons (β^+ decay), one must be careful to balance electron masses on the two sides of the decay equation. To compute the Q-value for the decay $^{40}_{20}\text{Ca} \rightarrow \text{e}^+ + ^{40}_{19}\text{K}$, we add 20 electrons to each side to create neutral atoms.

Then we have $^{40}_{20}\text{Ca}_\text{atom} \rightarrow e^- + e^+ + ^{40}_{19}\text{K}_\text{atom}$ and make use of the neutral atomic masses given in Appendix B of the textbook. This gives

$$Q = \left[m_{^{40}_{20}\text{Ca}_\text{atom}} - 2m_e - m_{^{40}_{19}\text{K}_\text{atom}} \right] c^2$$

$$= \left[39.962\,591 \text{ u} - 2(0.000\,549 \text{ u}) - 39.963\,999 \text{ u} \right](931.5 \text{ MeV/u})$$

$$= -2.33 \text{ MeV}$$

Since $Q < 0$, this decay cannot occur spontaneously. ◊

(b) For the decay $^{144}_{60}\text{Nd} \rightarrow ^4_2\text{He} + ^{140}_{58}\text{Ce}$, we add 60 electrons to each side and obtain $^{144}_{60}\text{Nd}_\text{atom} \rightarrow ^4_2\text{He}_\text{atom} + ^{140}_{58}\text{Ce}_\text{atom}$. Again using neutral atomic masses from Appendix B,

$$Q = \left[m_{^{144}_{60}\text{Nd}_\text{atom}} - m_{^4_2\text{He}_\text{atom}} - m_{^{140}_{58}\text{Ce}_\text{atom}} \right] c^2$$

$$= \left[143.910\,083 \text{ u} - 4.002\,603 \text{ u}) - 139.905\,434 \text{ u} \right](931.5 \text{ MeV/u})$$

$$= +1.91 \text{ MeV}$$

In this case, $Q > 0$ and the decay may occur spontaneously. ◊

---

31. A wooden artifact is found in an ancient tomb. Its carbon-14 $\left(^{14}_6\text{C} \right)$ activity is measured to be 60.0% of that in a fresh sample of wood from the same region. Assuming the same amount of ^{14}C was initially present in the wood from which the artifact was made, determine the age of the artifact.

Solution

The activity of a radioactive sample is directly proportional to the number of radioactive nuclei contained in that sample. Thus, if the carbon-14 activity of the artifact is 60.0% that of an equal amount of fresh wood, the artifact must contain 60.0% of the number of carbon-14 nuclei found in the fresh sample. If it is assumed that the artifact originally contained the same number of ^{14}C nuclei as is found in a equal quantity of fresh wood, then we have

$$N = 0.600 N_0$$

or, 60.0% of the ^{14}C nuclei originally in the wood of the artifact are still present in that sample.

Since $R = \lambda N$, we then have $R = \lambda \left(0.600 N_0 \right) = 0.600 \left(\lambda N_0 \right) = 0.600 R_0$

where R_0 is the activity the artifact had when its wood was freshly cut.

But $R = R_0 e^{-\lambda t} = R_0 e^{-\left(\ln 2 / T_{1/2} \right) t}$

which gives $0.600 R_0 = R_0 e^{-(\ln 2 / T_{1/2}) t}$ or $\ln(0.600) = -\left(\dfrac{\ln 2}{T_{1/2}}\right) t$

The half-life of ^{14}C is $T_{1/2} = 5\ 730$ yr, so the time that has passed since the wood in the artifact was part of a living tree is

$$t = -\left(\frac{\ln 0.600}{\ln 2}\right) T_{1/2} = -\left(\frac{\ln 0.600}{\ln 2}\right)(5\ 730\ \text{yr}) = 4.22 \times 10^3\ \text{yr}$$ ◊

33. Identify the unknown particles X and X′ in the following nuclear reactions:

$$X + {}_2^4 He \rightarrow {}_{12}^{24} Mg + {}_0^1 n$$

$$_{92}^{235} U + {}_0^1 n \rightarrow {}_{38}^{90} Sr + X + 2 {}_0^1 n$$

$$2 {}_1^1 H \rightarrow {}_1^2 H + X + X'$$

Solution

In any reaction, both the total number of nucleons and the total charge must be the same after the reaction as it was before. We use these criteria to identify the unknown particles.

In the reaction $_Z^A X + {}_2^4 He \rightarrow {}_{12}^{24} Mg + {}_0^1 n$, we must have $A + 4 = 24 + 1$, giving $A = 21$. Also, it is necessary that $Z + 2 = 12 + 0$, or $Z = 10$. Thus, the unknown particle has a charge of $Z = +10e$ and contains a total of $A = 21$ nucleons. This is the nucleus $_{10}^{21} Ne$.   ◊

For the reaction $_{92}^{235} U + {}_0^1 n \rightarrow {}_{38}^{90} Sr + {}_Z^A X + 2 {}_0^1 n$, we require that $235 + 1 = 90 + A + 2(1)$, yielding $A = 144$. We also require that $92 + 0 = 38 + Z + 0$, or $Z = 54$. Therefore, the unknown particle must be the nucleus $_{54}^{144} Xe$.   ◊

Finally, for the proton-proton reaction, $2 {}_1^1 H \rightarrow {}_1^2 H + {}_Z^A X + X'$, we assume the particle X′ is probably a chargeless, massless particle such as a photon or neutrino. Then, it is necessary that $2(1) = 2 + A + 0$, which gives $A = 0$. Also, it is necessary that $2(1) = 1 + Z + 0$, or $Z = +1$. Thus, the unknown particle X must be a non-nuclear particle with a charge of $+1e$. This is the positron, β^+ or $_{+1}^0 e$. The particle X′ that accompanies the production of a positron is the neutrino, ν_e. Chapter 30 will explain the reason for this when the requirement to conserve electron-lepton number is discussed.   ◊

37. Natural gold has only one isotope: $^{197}_{79}\text{Au}$. If gold is bombarded with slow neutrons, e^- particles are emitted. (a) Write the appropriate reaction equation. (b) Calculate the maximum energy of the emitted beta particles. The mass of $^{198}_{80}\text{Hg}$ is 197.966 75 u.

Solution

(a) When $^{197}_{79}\text{Au}$ is bombarded by slow neutrons, it is possible for the gold nucleus to capture the neutron, momentarily forming an highly unstable $^{198}_{79}\text{Au}$ nucleus, which quickly undergoes beta decay, emitting an e^- particle and an antineutrino, $\bar{v}$. This process is summarized by the reaction equation:

$$^{197}_{79}\text{Au} + ^{1}_{0}\text{n} \rightarrow ^{198}_{80}\text{Hg} + ^{0}_{-1}\text{e} + \bar{v}_e$$ ◊

(b) To compute the Q-value for this reaction, we add 79 electrons to each side to form neutral atoms, yielding $^{197}_{79}\text{Au}_{\text{atom}} + ^{1}_{0}\text{n} \rightarrow ^{198}_{80}\text{Hg}_{\text{atom}} + \bar{v}_e$, and use neutral atomic masses from Appendix B of the textbook to obtain

$$Q = \left[m_{^{197}_{79}\text{Au}_{\text{atom}}} + m_n - m_{^{198}_{80}\text{Hg}_{\text{atom}}} - m_{\bar{v}_e} \right]c^2$$

$$= [196.966\ 552\ \text{u} + 1.008\ 665\ \text{u} - 197.966\ 75\ \text{u} - 0](931.5\ \text{MeV/u})$$

$$= +7.89\ \text{MeV}$$

Neglecting any recoil kinetic energy of the product nucleus, $^{198}_{80}\text{Hg}$, the 7.89 MeV of energy released in the reaction may be divided between the kinetic energies of the beta particle and the antineutrino in any proportions. Thus, the kinetic energies of the beta particles could range from a low of zero to a maximum of 7.89 MeV. ◊

---

43. A "clever" technician decides to heat some water for his coffee with an x-ray machine. If the machine produces 10 rad/s, how long will it take to raise the temperature of a cup of water by 50°C. Ignore heat losses during this time.

Solution

The radiation absorbed dose (or rad for short) is that amount of radiation which deposits 10^{-2} J of energy into 1 kg of absorbing material. Therefore, if the x-ray machine produces 10 rad/s, it would add energy to a mass m of absorbing material at a rate of

$$\frac{\Delta Q}{\Delta t} = m\left[\left(10\ \frac{\text{rad}}{\text{s}}\right)\left(\frac{10^{-2}\ \text{J/kg}}{1\ \text{rad}}\right)\right] = m(0.100\ \text{J/kg}\cdot\text{s})$$

The energy required to raise the temperature of a quantity of water having mass m by 50°C is

$$Q = mc(\Delta T) = m(4\ 186\ \text{J/kg}\cdot°\text{C})(50°\text{C}) = m(2.1\times10^5\ \text{J/kg})$$

The amount of time the x-ray machine will require to add this quantity of energy to the "clever" technician's coffee water is

$$\Delta t = \frac{Q}{\Delta Q/\Delta t} = \frac{m(2.1\times 10^5 \text{ J/kg})}{m(0.100 \text{ J/kg}\cdot\text{s})} = 2.1\times 10^6 \text{ s}\left(\frac{1 \text{ d}}{86\,400 \text{ s}}\right) = 24 \text{ d} \qquad \Diamond$$

---

49. A 200.0-mCi sample of a radioactive isotope is purchased by a medical supply house. If the sample has a half-life of 14.0 days, how long will it keep before its activity is reduced to 20.0 mCi?

Solution

If a radioactive sample has an activity R_0 at time $t = 0$, its activity at any time t later is given by

$$R = R_0 e^{-\lambda t}$$

where the decay constant λ is related to the half-life by the expression

$$\lambda = \frac{\ln 2}{T_{1/2}}$$

Thus, if the sample had an initial activity of $R_0 = 200.0$ mCi, we find the time when the activity will be $R = 20.0$ mCi $= 0.100\,R_0$ from

$$0.100 R_0 = R_0 e^{-\lambda t} \qquad \text{or} \qquad e^{\lambda t} = 10.0$$

Taking the natural logarithm of both sides of the last result gives

$$\lambda t = \ln(10.0) \qquad \text{or} \qquad t = \frac{\ln(10.0)}{\lambda} = \left[\frac{\ln(10.0)}{\ln 2}\right]T_{1/2}$$

Thus, if the sample has a half-life of $T_{1/2} = 14.0$ d,

$$t = \left[\frac{\ln(10.0)}{\ln 2}\right](14.0 \text{ d}) = 46.5 \text{ d} \qquad \Diamond$$

53. A medical laboratory stock solution is prepared with an initial activity due to ^{24}Na of 2.5 mCi/ml, and 10.0 ml of the stock solution is diluted at $t = 0$ to a working solution whose total volume is 250 ml. After 48 h, a 5.0-ml sample of the working solution is monitored with a counter. What is the measured activity? *Note:* 1 ml = 1 milliliter.

Solution

At $t = 0$, the total activity of the working solution is

$$R_0 = (2.5 \text{ mCi/ml})(10 \text{ ml}) = 25 \text{ mCi}$$

This activity is contained in a solution having a total volume of $V_{ws} = 250$ ml. Thus, the initial concentration of the activity is

$$c_0 = \frac{R_0}{V_{ws}} = \frac{25 \text{ mCi}}{250 \text{ ml}} = 0.10 \text{ mCi/ml}$$

The initial activity of the material that will make up the 5.0-ml sample was

$$R_{0,\text{sample}} = c_0 V_{\text{sample}} = (0.10 \text{ mCi/ml})(5.0 \text{ ml}) = 0.50 \text{ mCi}$$

The half-life of ^{24}Na is $T_{1/2} = 14.96$ h, so the decay constant is

$$\lambda = \frac{\ln 2}{T_{1/2}} = \frac{\ln 2}{14.96 \text{ h}}$$

The activity of the 5.0-ml sample at $t = 48$ h is then

$$R_{\text{sample}} = R_{0,\text{sample}} e^{-\lambda t} = (0.50 \text{ mCi}) e^{-\left(\frac{\ln 2}{14.96 \text{ h}}\right)(48 \text{ h})} = 5.4 \times 10^{-2} \text{ mCi}$$

or

$$R_{\text{sample}} = (5.4 \times 10^{-2}) \times 10^{-3} \text{ Ci} = 5.4 \times 10^{-5} \text{ Ci} = 54 \times 10^{-6} \text{ Ci} = 54 \text{ } \mu\text{Ci} \qquad \lozenge$$

30

Nuclear Energy and Elementary Particles

NOTES FROM SELECTED CHAPTER SECTIONS

30.1   Nuclear Fission

Nuclear fission occurs when a heavy nucleus, such as ^{235}U, splits (fissions) into two smaller nuclei. In such a reaction, *the total rest mass of the products is less than the original rest mass.* The sequence of events in the fission process is:

- The ^{235}U nucleus captures a thermal (low energy) neutron.

- This capture results in the formation of ^{236}U* (* indicates excited state), and the excess energy of this nucleus causes it to undergo violent oscillations.

- The ^{236}U* nucleus becomes highly elongated, and the force of repulsion between protons in the two halves of the dumbbell-shaped nucleus tends to increase the distortion.

- The nucleus splits into two fragments, emitting several neutrons in the process.

A nuclear reactor is a system designed to maintain a **self-sustained chain reaction.** The **reproduction constant K** is defined as the average number of neutrons released from each fission event that will cause another event. In a power reactor, it is necessary to maintain a value of K close to 1. Under this condition, the reactor is said to be **critical.**

Important factors and processes relative to the design and operation of a nuclear reactor include:

- **Neutron leakage** — Some neutrons produced by a fission event will escape the volume of the reactor core before producing another fission event. The fraction of neutrons lost by leakage increases as the ratio of the surface area to the volume of the reactor vessel increases.

- **Regulating neutron energy** — The fast neutrons produced by fission are reduced in energy by allowing them to undergo collisions with a material of low atomic number (moderator). Slow neutrons have a greater probability than fast neutrons of initiating subsequent fission events.

- **Non-fissioning nuclei** — Some neutrons are captured by nuclei that do not undergo fission. The probability that a capture event will occur is reduced when the neutron energies are lower.

- **Power level** — The power level of the reactor is controlled by adjusting the number and position of control rods made of materials that are efficient neutron absorbers.

- **Reactor safety** — The primary operational concerns related to reactor safety are: containment of the fuel and radioactive fission products, flow of coolant to the core, disposal of spent fuel rods, and transportation of fuel and waste products.

30.2   Nuclear Fusion

Nuclear fusion is a process in which two light nuclei combine to form a heavier nucleus. The mass of the heavy nucleus is less than the sum of the masses of the two light nuclei; this loss of mass is the source of energy released in a fusion event. The major obstacle in obtaining useful energy from fusion is the large Coulomb repulsive force between charged nuclei as they approach close separation. Sufficient energy must be supplied to the particles to overcome this Coulomb barrier and thereby enable the nuclear attractive force to take over. Important considerations in the effort to achieve controlled fusion are:

- **Ignition temperature** — temperature at which hydrogen nuclei have sufficient kinetic energy to overcome the Coulomb force of repulsion.

- **Ion density** — number of electrons and positive ions per cm^3 which are produced in the gas (plasma) at high temperature.

- **Confinement time** — time that the interacting ions in the plasma are maintained at a temperature high enough for the fusion reaction to proceed.

- **Lawson's criterion** — condition on plasma density and confinement time under which net power output is possible.

30.3   Elementary Particles and the Fundamental Forces

There are four fundamental forces in nature:

- The **strong force** is responsible for binding quarks to form neutrons and protons and is the nuclear force that binds neutrons and protons into nuclei. *It is a very short-range force and is negligible for separations greater than the approximate size of the nucleus.*

- The **electromagnetic force** is responsible for the binding of atoms and molecules. *It is a long-range force that decreases in strength as the inverse square of the separation between interacting particles.*

- The **weak force** is a short-range nuclear force that tends to produce instability in certain nuclei. The weak interaction governs the structure of basic matter particles and is responsible for beta decay. *The electromagnetic and weak forces are now believed to be manifestations of a single force called the* **electroweak force.**

- The **gravitational force** is the weakest of all the fundamental forces. It is a long-range force that holds the planets, stars, and galaxies together. *The effect of the gravitational force on elementary particles is negligible.*

We model the universe as composed of field particles and matter particles. The exchange of field particles mediates the interactions of the matter particles.

- **Gluons** are the field particles for the strong force.

- **Photons** are exchanged by charged particles in electromagnetic interactions.

- **W^+, W^-, and Z bosons** mediate the weak force.

- **Gravitons** are (as yet undetected) quantum particles of the gravitational field.

30.4 Positrons and Other Antiparticles

A particle and its antiparticle (for example, an electron and a positron) have the same mass but opposite charge. Particle-antiparticle pairs may have other opposite values such as lepton number and baryon number.

Pair annihilation is an event in which a particle-antiparticle pair initially at rest, combine with each other, disappear, and produce two photons which move in opposite directions with equal energy and with the same magnitude of momentum.

30.5 Classification of Particles

All particles (other than photons) can be classified into two categories: **hadrons** and **leptons.**

Hadrons are composed of quarks and interact via all fundamental forces. They include two classes of particles grouped according to their masses and spins:

- **Mesons** — decay finally into electrons, positrons, neutrinos, and photons; they have spin quantum number 0 or 1.

- **Baryons** — (with the exception of the proton) decay in a manner leading to end products which include a proton; they have spin quantum number $\frac{1}{2}$ *or* $\frac{3}{2}$.

Leptons (e^-, μ^-, τ^-, ν_e, ν_μ, ν_τ, and their antiparticles) interact by the weak interaction, the electromagnetic interaction (if charged), and (presumably) the gravitational interaction but not by the strong force. Leptons have spin number $\frac{1}{2}$. Electrons, muons, and neutrinos are included in this group.

30.6 Conservation Laws

Conservation laws in the study of elementary particles are based on empirical evidence:

- **Conservation of baryon number** — Whenever a nuclear reaction or decay occurs, the sum of the baryon numbers before the process must equal the sum of the baryon numbers after the process. This requirement is quantified by assiging a baryon number, B ($B = +1$ for baryons, $B = -1$ for antibaryons, $B = 0$ for all other particles).

- **Conservation of lepton number** — Conservation laws are quantified by assigning lepton numbers, one for each variety of lepton: electron-lepton number (L_e), muon-lepton number (L_μ), and tau-lepton number (L_τ). *In each case the sum of the lepton numbers before a reaction or decay must equal the sum of the lepton numbers after the reaction or decay.*

- **Conservation of strangeness** — Strange particles are always produced in pairs by the strong interaction and decay very slowly, as is characteristic of the weak interaction. Strange particles are assigned a **strangeness quantum number, S**; quantitatively, $S = \pm 1, \pm 2, \pm 3$ for strange particles and $S = 0$ for nonstrange particles. Whenever a nuclear reaction or decay occurs via the strong interaction or the electromagnetic interaction, the sum of the strangeness numbers before the process must equal the sum of the strangeness numbers after the process ($\Delta S = 0$). Decays occurring via the weak interaction include the loss of one strange particle. This process proceeds slowly and violates the law of conservation of strangness.

30.7   The Eightfold Way
30.8   Quarks and Color

Quarks and antiquarks (of six possible types or flavors) make up baryons and mesons; the identity of each particle is determined by the particular combination of quarks. *Baryons consist of three quarks, antibaryons consist of three antiquarks, and mesons consist one quark and one antiquark.* The table at right lists the charge and baryon number for each of the six quarks and antiquarks.

Quark, Antiquark	Charge Number	Baryon Number
Up: $u, \bar{u}$	$+\frac{2}{3}e, -\frac{2}{3}e$	$\frac{1}{3}, -\frac{1}{3}$
Down: $d, \bar{d}$	$-\frac{1}{3}e, +\frac{1}{3}e$	$\frac{1}{3}, -\frac{1}{3}$
Strange: $s, \bar{s}$	$-\frac{1}{3}e, +\frac{1}{3}e$	$\frac{1}{3}, -\frac{1}{3}$
Charmed: $c, \bar{c}$	$+\frac{2}{3}e, -\frac{2}{3}e$	$\frac{1}{3}, -\frac{1}{3}$
Bottom: $b, \bar{b}$	$-\frac{1}{3}e, +\frac{1}{3}e$	$\frac{1}{3}, -\frac{1}{3}$
Top: $t, \bar{t}$	$+\frac{2}{3}e, -\frac{2}{3}e$	$\frac{1}{3}, -\frac{1}{3}$

The strong force between quarks is referred to as the **color force** and is mediated by massless particles called **gluons. Color charge** (red, green, blue) is a property assigned to quarks, which allows combinations of quarks to satisfy the exclusion principle.

EQUATIONS AND CONCEPTS

The **fission of an uranium nucleus** by bombardment with a low energy neutron results in the production of **fission fragments** and typically two or three neutrons. The energy released in the fission event appears in the form of kinetic energy of the fission fragments and the neutrons.

$$^{1}_{0}n + ^{235}_{92}U \rightarrow ^{236}_{92}U^* \qquad (30.1)$$
$$\rightarrow X + Y + \text{neutrons}$$

$^{238}_{92}U^*$ is a short-lived intermediate state.

Mass and energy conservation requirements are satisfied by many different combinations of fission fragments. On the average, 2.47 neutrons are produced per fission event. Equation (30.2) shows a typical fission reaction.

$$_0^1 n + _{92}^{235}U \rightarrow _{56}^{141}Ba + _{36}^{92}Kr + 3\,_0^1 n \qquad (30.2)$$

The **fusion reactions** shown here are those most likely to be used as the basis for the design and operation of a fusion power reactor. The Q values refer to the energy released in each reaction. In the equations at right, D represents deuterium and T represents tritium.

$$_1^2 D + _1^2 D \rightarrow _2^3 He + _0^1 n \quad (Q = 3.27 \text{ MeV})$$

$$_1^2 D + _1^2 D \rightarrow _1^3 T + _1^1 H \quad (Q = 4.03 \text{ MeV})$$
$$(30.4)$$

$$_1^2 D + _1^3 T \rightarrow _2^4 He + _0^1 n \quad (Q = 17.59 \text{ MeV})$$

Lawson's criterion states the conditions under which a net power output of a fusion reactor is possible. In these expressions, n is the **plasma density** (number of ions per cubic cm), and τ is the plasma **confinement time** (the time during which the interacting ions are maintained at a temperature equal to or greater than that required for the reaction to proceed).

$$n\tau \geq 10^{14} \text{ s/cm}^3 \quad \text{(D-T reaction)}$$
$$(30.5)$$

$$n\tau \geq 10^{16} \text{ s/cm}^3 \quad \text{(D-D reaction)}$$

Three varieties of pions correspond to three charge states: π^+, π^-, and π^0. Pions are very unstable particles; the equations at right show a decay mode for each type of pion.

$$\pi^+ \rightarrow \mu^+ + \nu_\mu$$

$$\pi^- \rightarrow \mu^- + \bar{\nu}_\mu \qquad (30.6)$$

$$\pi^0 \rightarrow \gamma + \gamma$$

Muons are in two varieties: μ^- and μ^+ (the antiparticle). Example decay processes are shown in the equations at right.

$$\mu^+ \rightarrow e^+ + \nu_e + \bar{\nu}_\mu$$

$$\mu^- \rightarrow e^- + \nu_e + \bar{\nu}_\mu$$

REVIEW CHECKLIST

- Describe the sequence of events which occur during the fission process. Write an equation which represents a typical fission event. Use data obtained from the binding energy curve to estimate the disintegration energy of a typical fission event. (Section 30.1)

- List the major parameters influencing fission reactor design and operation which are important in maintaining a steady power level. (Section 30.1)

- Describe the basis of energy release in fusion and write out several nuclear reactions which might be used in a fusion-powered reactor. (Section 30.2)

- Calculate the minimum photon energy required to produce a given particle-antiparticle pair. (Section 30.4)

- Outline the classification of elementary particles and mention several characteristics of each group (relative mass value, spin, decay mode). (Section 30.5)

- Determine whether or not a suggested decay can occur based on the conservation of baryon number and the conservation of lepton number. (Section 30.6)

- Use conservation laws to identify reactants and products in a proposed reaction. (Section 30.6)

SOLUTIONS TO SELECTED END-OF-CHAPTER PROBLEMS

5.   Assume that ordinary soil contains natural uranium in amounts of 1 part per million by mass. (a) How much uranium is in the top 1.00 meter of soil on a 1-acre $\left(43\,560\text{ ft}^2\right)$ plot of ground, assuming the specific gravity of soil is 4.00?   (b) How much of the isotope ^{235}U, appropriate for nuclear reactor fuel, is in this soil? *Hint*: See Appendix B for the percent abundance of $^{235}_{92}$U.

Solution

(a)   The volume of soil in the top 1.00 meter on a one acre plot is

$$V = Ah = \left[43\,560\text{ ft}^2\left(\frac{1\text{ m}}{3.281\text{ ft}}\right)^2\right](1.00\text{ m}) = 4.05 \times 10^3\text{ m}^3$$

With a specific gravity of 4.00, this soil is four times as dense as water. Thus, the mass of soil in the top 1.00 m on a one acre plot is

$$m = \rho V = \left(4.00\rho_{\text{water}}\right)V$$

$$= 4.00\left(1.00 \times 10^3\text{ kg/m}^3\right)\left(4.05 \times 10^3\text{ m}^3\right) = 1.62 \times 10^7\text{ kg}$$

Since natural uranium makes up one part per million by mass of ordinary soil, the mass of uranium in this soil is

$$m_{uranium} = \frac{m}{1.00 \times 10^6} = \frac{1.62 \times 10^7 \text{ kg}}{1.00 \times 10^6} = 16.2 \text{ kg}$$ ◊

(b) The percentage abundance of the isotope ^{235}U in naturally occurring uranium is 0.720%. Thus, the mass of this isotope found in the 16.2 kg of natural uranium from the soil in the upper 1.00 m on a one acre plot is

$$m_{^{235}U} = \left(\frac{0.720}{100}\right) m_{uranium} = \left(\frac{0.720}{100}\right)(16.2 \text{ kg}) = 0.117 \text{ kg} = 117 \text{ g}$$ ◊

---

11. When a star has exhausted its hydrogen fuel, it may fuse other nuclear fuels. At temperatures above 1.0×10^8 K, helium fusion can occur. Write the equations for the following processes: (a) Two alpha particles fuse to produce a nucleus A and a gamma ray. What is nucleus A? (b) Nucleus A absorbs an alpha particle to produce a nucleus B and a gamma ray. What is nucleus B? (c) Find the total energy released in the reactions given in (a) and (b). *Note*: The mass of $^{8}_{4}$Be $= 8.005\ 305$ u.

Solution

(a) The first reaction in this sequence is where $^{A}_{Z}X$ is the unknown product nucleus A.

$$^{4}_{2}\text{He} + ^{4}_{2}\text{He} \rightarrow ^{A}_{Z}X + \gamma$$

Requiring that charge be conserved gives or $Z = 4$, meaning that A is an isotope of beryllium.

$$2 + 2 = Z + 0$$

Conserving the number of nucleons gives or $A = 8$. Thus, the product nucleus is $^{8}_{4}$Be.

$$4 + 4 = A + 0$$ ◊

(b) The next reaction in the sequence is where $^{A}_{Z}X$ is now the unknown product nucleus B.

$$^{8}_{4}\text{Be} + ^{4}_{2}\text{He} \rightarrow ^{A}_{Z}X + \gamma$$

Requiring that charge be conserved gives or $Z = 6$, meaning that B is an isotope of carbon.

$$4 + 2 = Z + 0$$

Conserving the number of nucleons gives or $A = 12$. Thus, the product nucleus is $^{12}_{6}$C.

$$8 + 4 = A + 0$$ ◊

(c) The net result of this pair of reactions is to fuse three alpha particles into a carbon-12 nucleus along with the production of two gamma rays. The overall mass deficit is

$$\Delta m = 3m_{^{4}_{2}\text{He}} - m_{^{12}_{6}\text{C}} = 3(4.002\ 603 \text{ u}) - 12.000\ 000 \text{ u} = 0.007\ 809 \text{ u}$$

The total energy released in this pair of reactions is then

$$Q = (\Delta m)c^2 = (0.007\ 809 \text{ u})(931.5 \text{ MeV/u}) = 7.274 \text{ MeV}$$ ◊

15. Assume a deuteron and a triton are at rest when they fuse according to the reaction

$$_1^2H + {}_1^3H \rightarrow {}_2^4He + {}_0^1n + 17.6 \text{ MeV}$$

Neglecting relativistic corrections, determine the kinetic energy acquired by the neutron.

Solution

Since both the deuteron and the triton are at rest before the reaction, the initial total momentum is zero. Thus, the total momentum must be zero after the reaction, meaning the alpha particle and the neutron must move in opposite directions with equal magnitude momenta, $p_\alpha = p_n$.

If we neglect relativistic corrections, the relation between the kinetic energy of a particle and its momentum is $KE = p^2/2m$, where m is the mass of the particle. Therefore, the kinetic energies of the alpha particle and neutron after reaction are related by

$$KE_\alpha = \frac{p_\alpha^2}{2m_\alpha} = \frac{p_n^2}{2m_\alpha} = \frac{2m_n KE_n}{2m_\alpha}$$

or

$$KE_\alpha = \left(\frac{m_n}{m_\alpha}\right) KE_n$$

The total energy converted from mass into kinetic energy during the reaction is given to be $Q = 17.6$ MeV. Since there was zero kinetic energy before reaction, the total kinetic energy after reaction must be

$$KE_n + KE_\alpha = Q = 17.6 \text{ MeV}$$

Thus, we have

$$\left(1 + \frac{m_n}{m_\alpha}\right) KE_n = 17.6 \text{ MeV}$$

or the kinetic energy acquired by the neutron during the reaction is

$$KE_n = \frac{17.6 \text{ MeV}}{\left(1 + m_n/m_\alpha\right)} = \frac{17.6 \text{ MeV}}{\left[1 + (1.008\ 665 \text{ u})/(4.002\ 603 \text{ u})\right]} = 14.1 \text{ MeV} \qquad \lozenge$$

18. A photon with an energy of 2.09 GeV creates a proton–antiproton pair in which the proton has a kinetic energy of 95.0 MeV. What is the kinetic energy of the antiproton?

Solution

The photon has zero rest energy, so the total energy present before the pair production process is $E_{\text{total}} = 2.09$ GeV $= 2\,090$ MeV.

Most of the total energy available will go into producing the rest energies of the particles created in this pair production process. The remainder will go into the kinetic energies of the pair of particles. From conservation of energy, we have

$$E_{\text{total}} = E_{R,p} + E_{R,\bar{p}} + KE_p + KE_{\bar{p}}$$

The proton and the antiproton have the same rest energy. From Table 30.2 in the textbook, this energy is found to be

$$E_{R,p} = E_{R,\bar{p}} = m_p c^2 = \left(938.3 \ \text{MeV}/c^2\right)c^2 = 938.3 \ \text{MeV}$$

Since it is specified that the kinetic energy of the proton is $KE_p = 95.0$ MeV, the kinetic energy of the antiproton must be

$$KE_{\bar{p}} = E_{\text{total}} - E_{R,p} - E_{R,\bar{p}} - KE_p$$

$$= 2\,090 \ \text{MeV} - 2\left(938.3 \ \text{MeV}\right) - 95.0 \ \text{MeV} = 118 \ \text{MeV} \qquad \lozenge$$

---

21. Each of the following reactions is forbidden. Determine a conservation law that is violated for each reaction.

(a) $p + \bar{p} \rightarrow \mu^+ + e^-$ (b) $\pi^- + p \rightarrow p + \pi^+$

(c) $p + p \rightarrow p + \pi^+$ (d) $p + p \rightarrow p + p + n$

(e) $\gamma + p \rightarrow n + \pi^0$

Solution

The basic conservation laws that must be observed in any particle reaction are: conservation of charge, Q; conservation of baryon number, B; conservation of lepton number (one for each variety of lepton), electron-lepton number L_e, muon-lepton number L_μ, and tau-lepton number L_τ; and conservation of strangeness. The strong and electromagnetic interactions always conserve strangeness. The weak interaction will violate conservation of strangeness, but by no more that one unit.

(a) The reaction $p + \bar{p} \rightarrow \mu^+ + e^-$ conserves charge, baryon number, and strangeness. It does not involve any tau-leptons, but does violate conservation of electron-lepton number ($L_{e,\,\text{before}} = 0 + 0$, $L_{e,\text{after}} = 0 + 1$) and conservation of muon-lepton number ($L_{\mu,\,\text{before}} = 0 + 0$, $L_{\mu,\text{after}} = -1 + 0$). $\qquad \lozenge$

(b) The reaction $\pi^- + p \rightarrow p + \pi^+$ conserves baryon number, strangeness, and all varieties of lepton number. However, it fails to conserve charge: $Q_{before} = -1 + 1 = 0$, $Q_{after} = +1 + 1 = 2$. ◊

(c) The reaction $p + p \rightarrow p + \pi^+$ conserves charge, strangeness, and all varieties of lepton number, but fails to conserve baryon number: $B_{before} = +1 + 1 = 2$, $B_{after} = +1 + 0 = 1$. ◊

(d) The reaction $p + p \rightarrow p + p + n$ conserves charge, strangeness, and all varieties of lepton number. However, it does not conserve baryon number: $B_{before} = +1 + 1 = 2$, $B_{after} = +1 + 1 + 1 = 3$. ◊

(e) The reaction $\gamma + p \rightarrow n + \pi^0$ conserves baryon number, strangeness, and all varieties of lepton number, but does not conserve charge: $Q_{before} = 0 + 1 = 1$, $Q_{after} = 0 + 0 = 0$. ◊

---

27. Find the number of electrons, and of each species of quark, in 1 L of water.

Solution

An ordinary water molecule (H_2O) contains two neutral atoms of ^1_1H and one neutral atom of $^{16}_8O$. Thus, each molecule contains 10 protons, 10 electrons, and 8 neutrons.

The molecular weight of water is 18.0 g/mol and its density is

$$\rho = 1.00 \times 10^3 \ \frac{kg}{m^3} \left(\frac{10^3 \ g}{1 \ kg} \right) \left(\frac{1 \ m^3}{10^3 \ L} \right) = 1.00 \times 10^3 \ g/L$$

Therefore, the mass of one liter of water is 1.00×10^3 g, and the number of molecules it contains is

$$N = \left(\frac{m}{M} \right) N_A = \left(\frac{1.00 \times 10^3 \ g}{18.0 \ g/mol} \right) \left(6.02 \times 10^{23} \ \frac{molecules}{mol} \right) = 3.34 \times 10^{25} \ molecules$$

The number of electrons, protons, and neutrons in the one liter of water is then

$$N_e = \left(10 \ \frac{electrons}{molecule} \right) \left(3.34 \times 10^{25} \ molecules \right) = 3.34 \times 10^{26} \ electrons \qquad ◊$$

$$N_p = \left(10 \ \frac{protons}{molecule} \right) \left(3.34 \times 10^{25} \ molecules \right) = 3.34 \times 10^{26} \ protons$$

and

$$N_n = \left(8 \ \frac{neutrons}{molecule} \right) \left(3.34 \times 10^{25} \ molecules \right) = 2.68 \times 10^{26} \ neutrons$$

Each proton (uud) contains 2 up quarks and 1 down quark, while each neutron (udd) has 1 up quark and 2 down quarks. Thus, the one liter of water will contain

$$N_u = 2N_p + N_n = 2(3.34 \times 10^{26}) + 2.68 \times 10^{26} = 9.36 \times 10^{26} \text{ up quarks} \qquad \Diamond$$

and

$$N_d = N_p + 2N_n = 3.34 \times 10^{26} + 2(2.68 \times 10^{26}) = 8.70 \times 10^{26} \text{ down quarks} \qquad \Diamond$$

---

31. A Σ^0 particle traveling through matter strikes a proton and a Σ^+ and a gamma ray, as well as a third particle, emerge. Use the quark model of each to determine the identity of the third particle.

Solution

The reaction of interest is $\Sigma^0 + p \rightarrow \Sigma^+ + \gamma + X$

where X is an unknown particle. We recognize that the Σ^0 particle and the proton are both baryons and interact via the strong interaction. Thus, we could attempt to identify the unknown particle by using the fact that the strong interaction always obeys several conservations laws: conservation of charge, conservation of baryon number, conservation of strangeness, and conservation of lepton numbers (one for each variety of lepton). However, we shall consider the quark composition (in terms of up, down, and strange quarks) of the known particles before and after the reaction to find the identity of particle X.

Consider Table 30.4 in the textbook to see that the quark composition of a Σ^0 particle is uds, that of a proton is uud, and that of a Σ^+ particle is uus. Thus, in terms of the quark composition, the reaction is

$$uds + uud \rightarrow uus + (0 \text{ quarks}) + (n_u n_d n_s)$$

where n_u, n_d, and n_s represents the number of up, down, and strange quarks, respectively, contained in the mystery particle X. Note that before the reaction, there are totals of 3 up quarks, 2 down quarks, and 1 strange quark present. After the reaction, we have: $n_u + 2$ up quarks, $n_d + 0$ down quarks, and $n_s + 1$ strange quarks present.

In order to conserve the net number of each type of quark in this reaction, we must have:

$$n_u + 2 = 3 \quad \text{or} \quad n_u = 1 \qquad \text{so } X \text{ must have 1 up quark}$$

$$n_d + 0 = 2 \quad \text{or} \quad n_d = 2 \qquad \text{so } X \text{ must have 2 down quarks}$$

and $\quad n_s + 1 = 1 \quad \text{or} \quad n_s = 0 \qquad \text{so } X \text{ must have 0 strange quarks}$

Thus, the quark composition of particle X must be udd and from Table 30.4 we see that particle X must be a neutron. $\qquad \Diamond$

---

33. Find the energy released in the fusion reaction

$$^1_1\text{H} + ^3_2\text{He} \rightarrow ^4_2\text{He} + e^+ + v_e$$

Solution

If we had nuclear masses (that is, masses of atoms stripped bare of all electrons), the equation for the energy released in the fusion reaction would be

$$Q = (\Delta m)c^2 = \left(m_{^1_1\text{H}} + m_{^3_2\text{He}} - m_{^4_2\text{He}} - m_e\right)c^2$$

where we have assumed that the neutrino has zero mass.

However, the masses given in Appendix B are those of neutral atoms. Thus, we must be very careful how we handle electron masses, especially in light of the fact that one of the product particles (the positron) in this reaction has the same mass as an electron.

To obtain neutral atoms, add three electrons to both sides of the reaction equation. We then have

$$^1_1\text{H}_\text{atom} + ^3_2\text{He}_\text{atom} \rightarrow ^4_2\text{He}_\text{atom} + e^- + e^+ + v_e$$

and the equation for the energy released becomes

$$Q = (\Delta m)c^2 = \left(m_{^1_1\text{H}_\text{atom}} + m_{^3_2\text{He}_\text{atom}} - m_{^4_2\text{He}_\text{atom}} - 2m_e\right)c^2$$

Using the masses of neutral atoms from Appendix B and the mass of the electron from the inside back cover of this manual gives the energy released in the fusion reaction as

$$Q = \left[1.007\ 825\ \text{u} + 3.016\ 029\ \text{u} - 4.002\ 603\ \text{u} - 2(0.000\ 549\ \text{u})\right]\left(931.5\ \frac{\text{MeV}}{\text{u}}\right)$$

$$Q = 18.8\ \text{MeV}$$ ◊

---

37. A 2.0-MeV neutron is emitted in a fission reactor. If it loses one half its kinetic energy in each collision with a moderator atom, how many collisions must it undergo in order to reach an energy associated with a gas at a room temperature of 20.0°C?

Solution

From the kinetic theory of gases (see Chapter 10 in the textbook), the average kinetic energy of a particle in a gas at room temperature ($T = 20.0°\text{C} = 293\ \text{K}$) is given by

$$KE_\text{av} = \frac{3}{2}k_B T = \frac{3}{2}(1.38 \times 10^{-23}\ \text{J/K})(293\ \text{K})\left(\frac{1\ \text{eV}}{1.60 \times 10^{-19}\ \text{J}}\right) = 0.0379\ \text{eV}$$

If the neutron with an initial kinetic energy of $KE_i = 2.0\ \text{MeV}$ loses one half of its incident kinetic energy in each collision with a moderator atom, its kinetic energy after n such collisions will be

$$KE_f = \left(\frac{1}{2}\right)^n KE_i = \frac{2.0\ \text{MeV}}{2^n}$$

Thus, if its final kinetic energy is to be $KE_f = KE_{av} = 0.0379$ eV, we must have 0.0379 eV $= (2.0$ MeV$)/2^n$, or $2^n = (2.0$ MeV$)/(0.0379$ eV$)$. To determine the number of collisions with moderator atoms required to make this true, we take the natural logarithm of both sides of this result, and obtain

$$n \ln 2 = \ln\left[\frac{2.0 \text{ MeV}}{0.0379 \text{ eV}}\left(\frac{10^6 \text{ ev}}{1 \text{ MeV}}\right)\right] = \ln\left(5.3 \times 10^7\right)$$

and

$$n = \frac{\ln\left(5.3 \times 10^7\right)}{\ln 2} = 26 \qquad \Diamond$$

---

39. (a) Show that about 1.0×10^{10} J would be released by the fusion of the deuterons in 1.0 gal of water. Note that 1 out of every 6 500 hydrogen atoms is a deuteron. (b) The average energy consumption rate of a person living in the United States is about 1.0×10^4 J/s (an average power of 10 kW). At this rate, how long would the energy needs of one person be supplied by the fusion of the deuterons in 1.0 gal of water? Assume that the energy released per deuteron is 1.64 MeV.

Solution

(a) The mass of 1.0 gal of water is

$$m = \rho V = \left(1.00 \times 10^3 \frac{\text{kg}}{\text{m}^3}\right)\left[1.0 \text{ gal}\left(\frac{3.786 \text{ L}}{1 \text{ gal}}\right)\left(\frac{10^{-3} \text{ m}^3}{1 \text{ L}}\right)\right] = 3.8 \text{ kg}$$

The number of molecules in this quantity of water is

$$N = n N_A = \left(\frac{m}{M}\right) N_A = \left(\frac{3.8 \times 10^3 \text{ g}}{18 \text{ g/mol}}\right)\left(6.02 \times 10^{23} \text{ molecules/mol}\right)$$

or $N = 1.3 \times 10^{26}$ molecules

Each of these water molecules contains two hydrogen nuclei, and one in every 6 500 of these is a deuteron. Thus, the number of deuterium nuclei $\left(^2_1\text{H}\right)$ in the gallon of water is

$$N_d = \frac{2N}{6\,500} = \frac{2\left(1.3 \times 10^{26}\right)}{6\,500} = 4.0 \times 10^{22} \text{ deuterons}$$

Assuming that the energy released per deuteron is 1.64 MeV, the total energy available from the 1.0 gal of water is

$$E = \left(4.0 \times 10^{22} \text{ deuterons}\right)\left(1.64 \frac{\text{MeV}}{\text{deuteron}}\right)\left(\frac{1.6 \times 10^{-13} \text{ J}}{1 \text{ MeV}}\right) = 1.0 \times 10^{10} \text{ J} \qquad \Diamond$$

(b)   The time this could supply the energy needs of one person is

$$t = \frac{E}{\text{consumption rate}} = \frac{1.0 \times 10^{10} \text{ J}}{1.0 \times 10^{4} \text{ J/s}} \left(\frac{1 \text{ d}}{86\ 400 \text{ s}} \right) = 12 \text{ d}$$ ◊

---

43.   The Sun radiates energy at the rate of 3.85×10^{26} W. Suppose the net reaction

$$4\text{p} + 2\text{e}^- \rightarrow \alpha + 2v_\text{e} + 6\gamma$$

accounts for all the energy released. Calculate the number of protons fused per second. *Note:* Recall that an alpha particle is a helium-4 nucleus.

Solution

To determine the energy released in each occurrence of the reaction $4\text{p} + 2\text{e}^- \rightarrow \alpha + 2v_\text{e} + 6\gamma$, we add two electrons to each side of the reaction equation to form neutral atoms and obtain

$$4\left({}_1^1\text{H}_\text{atom} \right) \rightarrow {}_2^4\text{He}_\text{atom} + 2v_\text{e} + 6\gamma$$

Then, using neutral atom masses from Appendix B of the textbook, and recognizing that both the neutrinos and the photons are massless particles, we find

$$Q = (\Delta m)c^2 = \left(4m_{{}_1^1\text{H}_\text{atom}} - m_{{}_2^4\text{He}_\text{atom}} \right)c^2$$
$$= \left[4(1.007\ 825 \text{ u}) - 4.002\ 603 \text{ u} \right](931.5 \text{ MeV/u})$$
$$= 26.7 \text{ MeV}$$

Each occurrence of this reaction consumes four protons. Thus, the energy released per proton consumed is $E_1 = 26.7 \text{ MeV}/4 \text{ protons} = 6.68 \text{ MeV/proton}$.

Therefore, the rate at which the Sun must be fusing protons to provide the power output that it has is

$$rate = \frac{\mathcal{P}}{E_1} = \frac{3.85 \times 10^{26} \text{ J/s}}{6.68 \text{ MeV/proton}} \left(\frac{1 \text{ MeV}}{1.60 \times 10^{-13} \text{ J}} \right) = 3.60 \times 10^{38} \text{ protons/s}$$ ◊